Walter Hower

**Discrete Mathematics**

## Also of Interest

*Elements of Discrete Mathematics*
*Numbers and Counting, Groups, Graphs, Orders and Lattices*
Volker Diekert, Manfred Kufleitner, Gerhard Rosenberger, Ulrich Hertrampf, 2024
ISBN 978-3-11-106069-9, e-ISBN 978-3-11-106255-6

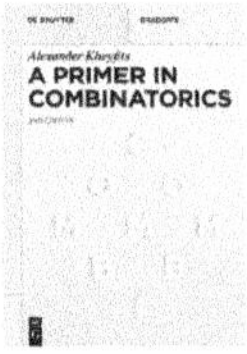


*A Primer in Combinatorics*
Alexander Kheyfits, 2021
ISBN 978-3-11-075117-8, e-ISBN 978-3-11-075118-5

*Geometry and Discrete Mathematics*
*A Selection of Highlights*
Benjamin Fine, Anja Moldenhauer, Gerhard Rosenberger, Annika Schürenberg, Leonard Wienke, 2022
ISBN 978-3-11-074077-6, e-ISBN 978-3-11-074078-3

*Quadratic Ideal Numbers*
*A Computational Method for Binary Quadratic Forms*
J. L. Lehman, 2025
ISBN 978-3-11-131935-3, e-ISBN 978-3-11-131936-0

*Linear Algebra*
*Vector and Inner Product Spaces*
Saurabh Chandra Maury, 2025
ISBN 978-3-11-151570-0, e-ISBN 978-3-11-151603-5

*Advanced Mathematics*
*An Invitation in Preparation for Graduate School*
Patrick Guidotti, 2022
ISBN 978-3-11-078085-7, e-ISBN 978-3-11-078092-5

Walter Hower

# Discrete Mathematics

Combinatorics, Counting, Proofs, Recurrences, Solutions

DE GRUYTER

**Author**
Prof. Dr. Dipl.-Inform. Walter Hower
Hochschule Albstadt-Sigmaringen
Informatik
Poststr. 6
D-72458 Albstadt-Ebingen
Germany
hower@hs-albsig.de

ISBN 978-3-11-120643-1
e-ISBN (PDF) 978-3-11-120689-9
e-ISBN (EPUB) 978-3-11-120724-7

**Library of Congress Control Number: 2025933823**

**Bibliographic information published by the Deutsche Nationalbibliothek**
The Deutsche Nationalbibliothek lists this publication in the Deutsche Nationalbibliografie; detailed bibliographic data are available on the Internet at http://dnb.dnb.de.


Cover image: Walter Hower
Typesetting: VTeX UAB, Lithuania

www.degruyter.com
Questions about General Product Safety Regulation:
productsafety@degruyterbrill.com

To my friends in Maths—and the brave ones who still want/have ;) to become one ⌣̈

# Opening

It is really an honor to present the translation of my corresponding German textbook cited at the end.

Some of my intended jokes were too difficult to change the language; therefore, I omitted these, and now may help to find an even larger audience.

Thanks to my wife and daughters for their understanding and mental support as well as to the publisher along with the *Sky*LaTeX team; furthermore, I am grateful to all the participants of my lectures, particularly my students, who addressed their comments to me.

Last but not least, I'm greatful to Ken Rosen, who gave me constructive feedback, including my initial English wording and hints regarding the appendix. (Of course, remaining formulations are always at the author's responsibility.)

Now it is your time—to hopefully enjoy this readable LaTeXt, which will impart knowledge with fun.

*WHo*

https://doi.org/10.1515/9783111206899-203

# Foreword

In this new book, Professor Walter Hower shows how Mathematics can be entertaining as well as compelling. He introduces important topics from Discrete Mathematics, the part of Mathematics dedicated to the study of discrete objects. Discrete Mathematics is needed to solve many important problems such as determining the probability of winning a lottery, building a circuit to add integers, counting the number of valid internet addresses, finding a link between two computers on a network, identifying spam emails, and countless other practical problems. This material is used throughout the mathematics sciences, including most aspects of Computer Science such as Artificial Intelligence, algorithm design, computer security, database theory, and automata theory. Students need this foundation before they study advanced courses in Computer Science and advanced courses in Mathematics that deal with Discrete Mathematics, including mathematical logic, set theory, number theory, linear algebra, abstract algebra, combinatorics, graph theory, and discrete probability theory.

Walter Hower has received awards recognizing him as an excellent teacher of Computer Science. He has great enthusiasm for the subject, as can be observed in his TED talk on Mathematics. He is noted for his excellent teaching style and the clarity and accessibility of his exposition. This book begins with key foundations, starting with the natural numbers, the concepts of functions and relations, and naive set theory. It continues with axiomatic set theory with a discussion of the generalized continuum hypothesis. It goes on to cover Boolean algebra and Boolean functions. Methods of proofs are then discussed, including direct and indirect proofs and the important proof method of mathematical induction. The book then moves on to combinatorics, which is the art of counting, staring with fundamental counting techniques based on permutations and combinations, and the famous pigeonhole principle. More advanced counting techniques based on recurrence relations come next, including introducing the Sterling number of the 1st and 2nd kinds and the Bell numbers. The book concludes with a chapter on discrete probability theory, addressing both general and conditional probability theory. An excellent example illustrating how Hower charms students is his treatment of the notorious Monty Hall problem that has long been the subject of heated discussion.

Students who learn the important aspects of Discrete Mathematics covered in this book will find that they have learned many essential tools that will help to make them successful in their future studies in the mathematics science, and especially in Computer Science. Their minds have been fertilized and they will also have experienced the joy of Mathematics.

Middletown, New Jersey, USA — Dr. Kenneth H. Rosen

https://doi.org/10.1515/9783111206899-204

# Contents

# Introduction

The present book roughly corresponds to the main part of a second-semester lecture—supported by about 222 (mainly interactively presented) course slides. Nevertheless, this book is of course self-contained, usable in a "stand-alone" mode—without interaction (in class or online).

Many students are stuck in *Discrete Mathematics*; this book should help to grasp the abstract formalism in a gentle way.

The structure of the booklet does not recapitulate some historical line or in a whatever typical order; it just serves the pragmatism needed. In some places, we already bring forward a certain notion, fully introduced later on, accompanied by common sense—which appears as a safe technique.

The topics selected are important for Computer Science students; hence, accuracy is an issue—which should get trained by our material here.

In the foundational first chapter, we start with the *natural numbers*, introduce *functions* and illustrate the realm of *relations*. In the second chapter, we lay the foundations of *set theory*, introduce further terms, present inherent laws and reason about the order of magnitude of finite as well as infinite sets. (This consideration of infinity prepares the ground for a more intuitive understanding of the uncomputability in Theoretical Computer Science.) The third chapter treats *Boolean algebra*. Terms and laws get presented and corresponding truth tables are set up; interesting to mention are the formulae for the number of possible assignments and functions. The fourth chapter harbors the common *proof principles*. We present *induction* on natural numbers and on strings as well as the *direct* and *indirect proof*. With the fifth chapter, we enter the roller coaster ⌣̈ of the *counting techniques*. The *rule of sum*, *product* and *quotient* can be found here as well as the *pigeonhole principle* and the mechanism of *in/exclusion*. The *recurrence relation*, with follows there, can be used as a tool of creativity, in order to be allowed to hope for a closed formula in a counting challenge. There are still solutions for *ordering* and *selection problems* to come, together with *permutation* and *binomial coefficient*. Additionally, as delicacy, I provide the *Stirling numbers of the first* and *second kind*, especially in their versions of *cycle* and *subset numbers* and finally the *Bell numbers*. The sixth chapter with the *general* and *conditional probability theory* with the obligatory "Monty Hall problem" provides the showdown. ⌣̈

It has become a rather handy copy; this should tempt you to use this book with pleasure. Despite its tight size, maybe exactly for this effort to formulate concisely and without unnecessary overhead, it has still been some piece of work.

Now it is done; have fun (at least intellectually ⌣̈) with
*Discrete Mathematics—Combinatorics, Counting, Proofs, Recurrences, Solutions*

https://doi.org/10.1515/9783111206899-001

# 1 Foundations

We start quite harmlessly with the basics of the field. The *Peano axioms* are surely fundamental along with the set of *natural numbers*—which immediately is the subject of the first section. In the second section, we discuss the notion of a *function* along with some special features. In the third section, we face the *relations*; this includes the general $n$-fold Cartesian product as well as the concept of the *lattice* with its *partial order*. This first chapter might suffice to get mentally "booted" ☺.

## 1.1 Basis

We introduce the *natural numbers* and illustrate the five Peano axioms.

By $\mathcal{N}$, we label the (infinitely large) set of *natural numbers* $\mathcal{N} :=^1 \{0, 1, 2, 3, \ldots\}$. In Computer Science, it is inevitable to use the "0" as smallest figure—the reason why we do not introduce $\mathcal{N}_0$ as label for this set. Once we like/need to exclude the "0," starting by "1," we define $\mathcal{N}^+ := \mathcal{N} \setminus \{0\}$ [= $\{1, 2, 3, \ldots\}$], sometimes denoted by $\mathcal{N}_1$. Here, now come the axioms for $\mathcal{N}$ according to Giuseppe Peano:

1. "0" is a natural number.
2. "0" is not a successor (:= $n + 1$) of a natural number ($n$).
3. Every natural number $n$ is a predecessor of exactly one successor number $n + 1$.
4. Different natural numbers have different successor numbers.
5. Once in an area $T$ ("$\subseteq$") of the natural numbers the "0" is included and in general for each number in $T$ also ("$\longrightarrow$") its successor number is in $T$ as well, then ("$\Longrightarrow$") $T = \mathcal{N}$.

More formally the *Peano axioms* look as follows:

1. $0 \in \mathcal{N}$; "$\in$" means: "(is) element of."
2. $0 \neq s(n) := n + 1, n \in \mathcal{N}$; $s$ := "successor" is the successor *function*.[2]
3. $\forall n \in \mathcal{N}\ \exists!\, s(n)$; $\forall$ := "for all" (All quantor), $\exists$ := "exists" (Existence qu.).[3]
4. $n_1 \neq n_2 \Longrightarrow s(n_1) \neq s(n_2)$; the successor function is *injective*.[4]
5. $0 \in T_{[\subseteq \mathcal{N}]}, \forall n_{[\in \mathcal{N}]} \in T \longrightarrow s(n) \in T \Longrightarrow T = \mathcal{N}$; "induction axiom."

The axiom mentioned last represents the basis of the proof principle of *induction*; see the general procedure of the induction step in Subsection 4.1.1 (p. 33).

---

1 "$l := r$" means that the *l*eft-hand side gets its value from the *r*ight-hand one, "$l =: r$" means that the *r*ight-hand side gets its value from the *l*eft-hand one; the value assignment always works toward the direction of the column.

2 *f.*: see subsequent Section 1.2.

3 "!" behind it (see above) means: "exactly 1".

4 See the following Section 1.2 from page 5.

https://doi.org/10.1515/9783111206899-002

Here now comes the moment to mention the unproven status of axioms (and also of so-called "hypotheses"): Axioms are (in general) not provable. They just get—but at least—considered to be reasonable; they are only not in contradiction to the current mathematical understanding. However, exactly at this place, this could run—at some time—into a severe problem. Once we would find s. th. being in contradiction to an axiom, we had to decide which attitude does make more sense. This could imply that we have to abandon the axiom. If this situation were to happen to the 5th Peano axiom, induction proofs could get deleted.[5]

## 1.2 Functions

Let us now come to the central term of the *function* along with some specializations:

$$f : D \to C.$$

It provides to every input element of the "start" set $D$ (left of the arrow) exactly 1 output element of the "target" set $C$ (right of the arrow)—notationally:

$$\forall x \in D \quad \exists! f(x) \in C.$$

Such a function is also called *total*.

For a *partial* function, this fact of the above mentioned "$\forall$" is not required. (The so-called "definition gaps" are allowed.)

Therefore, every (total) function also is a partial, albeit special, function:[6] $f$ total $\Longrightarrow f$ partial.

The converse is, of course, not true; i. e.:[7] $f$ partial $\not\Longrightarrow f$ total.

Similarly, the following holds: $f$ partial $\not\Longrightarrow f\neg$ ("not") total, as illustrated in footnote 6; clearly (although unimportant here): $f\neg$ partial $\Longrightarrow f\neg$ total.

Let us now illuminate the sets $D$ of inputs and $C$ of outputs:

The *domain* ($D$) represents all possibilities of the input into the function. (For each input case, the corresponding output value must clearly be defined.)

The (potential) *codomain* ($C$) represents the values, which might get used by the function for its output. (There is no need for each potential value to really get selected by the function.)

The *image* (also called *range*) ($B$[8]) finally represents exactly those values of the codomain, which really get produced by the corresponding function. (In the follow-up

5 In 2025, everything still was fine.

6 In this (special) case—totally defined—without definition gaps.

7 The following sign "$\not\Longrightarrow$" means "does not imply".

8 In German, "*B*ild-Menge".

chapter, we introduce for this simple set-theoretic connection that all elements of a set [here $B$] are completely included in another set [here $C$], the term "[improper] subset," by the following symbol: $B \subseteq C$.) The image $B$ must both be "correct" as well as "complete":[9] All possible values are used as outputs by the function, and no value needed by the function is missing.

Once we focus for a function $f$, only on a (usually "proper") subset $S$ of the domain $D$, we call this a *restriction*:

$$f_{|S}^{D\to C} : S_{[\subseteq D]} \to C.$$

After these preparatory notions, we now introduce some concrete functions:

- *i*nclusion

$$i : D \to C,$$
$$i(x) := x.$$

  In $C$, at least the elements from $D$ must get provided, because all $D$ elements might get used; $D$ is "included" in $C$.[10]

- *id*entity

$$id : S \to S,$$
$$id(x) = i(x)_{[C=D=:S]} := x.$$

  The *id*entity mapping is a special *i*-function, where the domain is identical both to the codomain and the image—the codomain does not even provide unnecessary values; hence, no need to work on a proper superset (cf. Section 2.2).

- *p*rojection

$$\pi_j : \mathop{\times}_{i:=1}^{n} D_i \to D_j,$$
$$\pi_j(x_1, x_2, x_3, \ldots, x_n) := x_j, \quad 1 \le j \le n.$$

  The projection typically has a higher-ary input structure,[11] e. g., a pair ("binary," triple ("ternary"), etc., up to an arbitrary $n$-tuple.
  Now we investigate the "cross-product" sign ("Cartesian product") representing the *d*omain $D$. For each of th $n$ variables, there exists an individual input set $D_i$, out of which the corresponding $x_i$ $(1 \le i \le n)$ can take whatever input value; which yields an $n$-ary input. Now, just one additional piece of information is missing to perform the projection: the selection position $j$ $(1 \le j \le n)$, which we focus on; this number

9 A well-known pair of notions.

10 $C$ is a superset of $D$ ($C \supseteq D$), $D$ is a "subset" of $C$ ($D \subseteq C$); cf. Section 2.2, p. 14.

11 If only 1 (=: $n$) input parameter, it is called "unary".

is also called the "index" (the parameter interested in). The actual operation takes place completely without any pain ◡̈; the projection just delivers the content of the parameter $j$ thereby selected: $x_j$.

- We also survive the following *injection* without anaesthetic ◡̈: ("1-to-1"), injective $function_{[D\to C]}$; $x_1 \neq x_2 \Longrightarrow f_{\texttt{injective}}(x_1) \neq f_{\texttt{injective}}(x_2)$.
  Hence, such a function yields for different input cases different unique output values; therefore, the following holds:[12] $|D| \leq_{f_{\texttt{injective}}} |C|$.
  (Later on, by the famous [and simple] "pigeonhole principle,"[13] we learn that for an injection $|D| \not> |C|$.)
- The other view regarding the magnitude of the *D*omain in comparison to the *C*odomain yields the *surjection* ("onto"), surjective $function_{[D\to C]}$; $\forall y \in C\ \exists x \in D : f(x) = y$.
  Such a function exploits the codomain completely—the image (set) $B$ equals the codomain $C$. Due to the (general) feature of a function, we then have $|D| \geq_{f\ \texttt{surjective}} |C|$, equivalent to $|D| \not< |C|$.
- Once the function is both injective as well as surjective, we have a: *bijection* ("1-to-1" correspondence), bijective $function_{[D\to C]}$.
  This yields to the identity of the order of magnitudes of *d*omain and *c*odomain: $|D| =_{f\ \texttt{bijective}} |C|$.
- The *inverse* $f^{-1}$ of a $function$ $f$ yields the source of a value formerly produced: inverse function.
  Due to the fact that the original $f$ as well as $f^{-1}$ itself are total functions, we know the following: On the one hand, only injective functions can get inverted (otherwise the inversion does not provide a unique value); on the other hand, also the inverse function must be totally defined, the reason why the original function must additionally be surjective. Hence, we have the following universally valid statement: Just bijections are invertible!
  Therefore, given a bijective function $f : X \to Y$; we notate: $f^{-1} : Y \to X, \forall y \in Y$ $\exists! f^{-1}(y) =: x \in X$ with $f(x) = y$.
  Due to the fact that $f$ and $f^{-1}$ are bijections, the corresponding $x$ is, of course, unique;[14] $x$ depends on $y$ and $f^{-1}$: for each $y$ there exists a different $x$, according to the bijective function. ($|X| = |Y|$.) $f_{[X\to Y]}$ is *invertible*: $\exists$ an inverse $f^{-1}_{[Y\to X]}$.
- Functions can get nested; this is called "composition":
  $h := g \circ f$, spoken: $g$ "after (execution of)" $f$.
  With indication of domain and codomain, it looks as follows:

$$f : X \to Y, \quad g : Y \to Z; \quad h := g \circ f : X \to Z;$$

12 $|\cdots|$ around a finite set $S$ means the number of (#) elements in it, in general "order of magnitude."

13 See Subsection 5.1.4, starting on page 48.

14 Generally, the following notation would be (for $|Y| > 1$) wrong: $\exists! x \in X\ \forall y \in Y : f(x) = y$.

and with input parameter, we have this nesting:

$$h(x) := g(f(x)) \in Z.$$

The input $x$ moves to the $f$-function, the output $f(x)$ produced thereby serves as input into the $g$-function, and the final output is the result of the composition. In general, the following holds: $g(f(x)) \neq f(g(x))$; illustration:[15]

$$D_f := C_f := D_g := C_g := \mathcal{N}; \qquad f(x) := x + 1, \quad g(x) := 2^x;$$
$$f(g(1)) := f(2^1) = f(2) := 2 + 1 = 3,$$
$$g(f(1)) := g(1 + 1) = g(2) := 2^2 = 4 \neq 3.$$

- Naturally, there exists also an *inverse composition*; $h^{-1} := (g \circ f)^{-1}$.
  Let us first construct the composition $h$ of the bijections $g$ and $f$: $h := g \circ f$. Due to the fact that $g$ gets executed after $f$, therefore $g$ gets computed last before the inversion, the inversion of the bijection $h$ by consecutive executions of the individual inversions are performed in this reverse order: first, $g^{-1}$, then $f^{-1}$; i. e., $f^{-1}$ after $g^{-1}$:

  $$h^{-1} := (g \circ f)^{-1} = f^{-1} \circ g^{-1}.$$

  By the indication of domain and codomain it becomes even clearer:

  $$f : X \to Y, \quad g : Y \to Z; \quad h := g \circ f : X \to Z;$$
  $$f^{-1} : Y \to X, \quad g^{-1} : Z \to Y; \quad h^{-1} := f^{-1} \circ g^{-1} : Z \to X.$$

  With input parameter, we get this nested notation:

  $$h^{-1}(z) := f^{-1}(g^{-1}(z)) \in X.$$

At the end, we still introduce three rounding functions, in order to yield from a positive real number a *natural* one:
- *floor*: $\lfloor x \rfloor$ := largest natural number $\leq x$;
- *ceiling*: $\lceil x \rceil$ := smallest natural number $\geq x$;
- *round*: $\lfloor x \rceil$ := choose($\lfloor x \rfloor, \lceil x \rceil$), in the (seldom) case it does not matter.

## 1.3 Relations

We discuss this term *relation* incl. some concretizations—whereby we favor the *binary* one (relation between 2 parameters):

$$a\,R\,b.$$

15 The fact that for some special cases (here for $x := 0$) "=" might be true, anyhow is irrelevant.

Here, the *r*elation $R$ is in the center of (the) 2 input values; hence, we talk about an "infix" notation. The following notation places $R$ as a "prefix," in front of the input:

$$R(a,b).$$

Each component of the pair arises from its own definition set:[16]

$$a \in A, \quad b \in B; \quad (a,b) \in A \times B =: C.$$

This describes the input structure. The "×" sign takes over the possibility (and here even the need) to select and combine every individual element from the sets (here $A$ and $B$) with each other. This is called the "cross-product" (or "*C*artesian product").[17] Is the $R$-property fulfilled, the corresponding pair gets added to this ($R$-)set of valid pairs. It is obvious that $C$ is a superset of $R$:[18] $C \supseteq R$.

Dealing, for the moment, with finite sets, we have

$$|C| = |A| \cdot |B| \geq |R|.$$

Example:[19]

$$A := \{0,1,2\}, \quad B := \{\underline{1},3\}; \quad C := \{0,1,2\} \times \{\underline{1},3\}, \quad R := <.$$
$$0,1 \in A, \ \underline{1} \in B; \quad (0,\underline{1}),\ (1,\underline{1}) \in A \times B =: C.$$
$$0 < \underline{1} : R \ni (0,\underline{1});\ (1,\underline{1}) \notin R \quad [1 \not< \underline{1}].$$
$$|R| = \big|\{(0,\underline{1}),(0,3),(1,3),(2,3)\}\big| = 4 \leq 6 = |C|.$$

This binary structure could get generalized to an arbitrary $n$-ary syntax:[20]

$$C := \mathop{\times}_{i:=1}^{n} S_i := S_1 \times S_2 \times S_3 \times \cdots \times S_n$$
$$:= \big\{(e_1,e_2,e_3,\ldots,e_n) \mid e_i \in S_i,\ 1 \leq i \leq n\big\},$$

which we could read as follows: set of all $n$-tuples $(e_1,\ldots,e_n)$, whereby each individual $e_i$ originates from its corresponding set $S_i$ $(1 \leq i \leq n)$:[21]

$$|C| = |S_1| \cdot |S_2| \cdot |S_3| \cdot \cdots \cdot |S_n| =: \prod_{i:=1}^{n} |S_i|.$$

---

16 Domain or codomain are not always called $D$ or $C$; $B$ need not be the name of the image (range).

17 Initially mentioned in Section 1.2 on page 4 as domain (of $n$-tuples) of the projection.

18 Nomenclature made precise in Section 2.2 starting on page 14.

19 The sign "∋", used here, means "contains".

20 "|" alone in a set usually means "such that" or "whereby," in the sense to further describe the set.

21 Every element has its exact position in the tuple; see additionally the preceding footnote 17.

Example ("$\hat{=}$" := "corresponds to"):

$$1 \le i \le n := 3, \quad S_i := \mathcal{B} := \{0,1\} \hat{=} \{false, true\};$$

$$|C| = \prod_{i:=1}^{3} |S_i| = |\mathcal{B}|^3 = 2^3 = 8.$$

Proof:

$$|C| = \left|\{(0,0,0),(0,0,1),(0,1,0),(0,1,1),(1,0,0),(1,0,1),(1,1,0),(1,1,1)\}\right|$$
$$= 8 = 2^3 = |\mathcal{B}|^3.$$

The following holds:

$$|\mathcal{B}^n| = |\mathcal{B}|^n = 2^n.$$

The proof proceeds according to Example 5 in Subsection 4.1.1 (p. 40).

Now, we introduce some notions of special binary *r*elations:

- reflexive: $a\,R\,a$
  example: $R$ := "as successful as"
- irreflexive: $a\,\not R\,a$
  ex.: $R$ := "is more stupid than"
- converse: $b\,R^{-1}\,a \Longleftrightarrow a\,R\,b$
  ex.: $R$ := "is half of"; $R^{-1}$ = "is the double of"
- complement: $a\,\bar{R}\,b \Longleftrightarrow a\,\not R\,b$
  ex.: $R$ := "="; $\bar{R}$ = "≠"
- composition: $a\,(R_2 \circ R_1)\,c \Longleftrightarrow a\,R_1\,b$ and[22] $b\,R_2\,c$
  ex.: $R_1$ := "twice as much", $R_2$ := "three times"; $R_2 \circ R_1$ = "6 times"
- symmetrical: $a\,R\,b \Longrightarrow b\,R\,a$
  ex.: $R$ := "sits close to"
- asymmetrical: $a\,R\,b \Longrightarrow b\,\not R\,a$
  ex.: $R$ := "is below"
- anti-symmetrical: $a\,R\,b$ and $b\,R\,a \Longrightarrow a = b$
  ex.: $R$ := "≥"
- transitive: $a\,R\,b$ and $b\,R\,c \Longrightarrow a\,R\,c$
  ex.: $R$ := ">"
- intransitive: $a\,R\,b$ and $b\,R\,c \Longrightarrow a\,\not R\,c$
  ex.: $R$ := "is exactly one place higher in the table"
- union: $a\,(R_1 \cup R_2)\,b \Longleftrightarrow a\,R_1\,b$ or[23] $a\,R_2\,b$
  ex.: $R_1$ := "notifies", $R_2$ := "visits"; $R_1 \cup R_2$ = "contacts"

22 Both cases must apply.

23 Including both cases together.

- intersection: $a\,(R_1 \cap R_2)\,b \Longleftrightarrow a\,R_1\,b$ and $a\,R_2\,b$
  ex.: $R_1 :=$ "≥", $R_2 :=$ "≠"; $R_1 \cap R_2 =$ ">"
- difference: $a\,(R_1 - R_2)\,b \Longleftrightarrow a\,R_1\,b$ and $a\,\not{R}_2\,b$
  ex.: $R_1 :=$ "≥", $R_2 :=$ "≠"; $R_1 - R_2 =$ "="
- symmetrical difference (e*X*clusive *OR*):[24]
  $a\,(R_1 \oplus R_2)\,b \Longleftrightarrow (a\,R_1\,b \wedge a\,\not{R}_2\,b) \vee (a\,\not{R}_1\,b \wedge a\,R_2\,b)$
  ex.: $R_1 :=$ "≥", $R_2 :=$ "≠"; $R_1 \oplus R_2 =$ "≤"
- pre-order: reflexive as well as transitive *r*elation
  ex.: $R :=$ "≥"
- equivalence relation: symmetrical pre-order
  ex.: $R :=$ "same meaning"
- partial order ⊑ (on a set $S$): anti-symmetrical preorder
  ex.: $R_{[\sqsubseteq]} :=$ "⊇".[25]

Based on this catalogue of terms, we apply the following concepts:
- partially ordered set, poSet$(S, \sqsubseteq)$: set $S$ with *partial order* ⊑
  ex.: $S :=$ set of all subsets of $\mathcal{B}$:[26] $\{\{\}, \{0\}, \{1\}, \{0,1\}\}$; $R_{[\sqsubseteq]} :=$ "⊇"
- comparable: 2 elements $a, b \in$ poSet are *comparable* $\Longleftrightarrow a \sqsubseteq b$ or $b \sqsubseteq a$
  ex.: $S := \{\{\}, \{0\}, \{1\}, \mathcal{B}\} =: 2^{\mathcal{B}}$, $R_{[\sqsubseteq]} :=$ "⊇" (see above), $a := \{0\}$, $b := \mathcal{B}$; $b \sqsubseteq a$
- incomparable: $a, b \in S$ are *incomparable* $\Longleftrightarrow a$ and $b$ are not comparable
  ex.: $S$, $R_{[\sqsubseteq]}$ and $a$ defined as above, $b := \{1\}$; $a \not\sqsubseteq b$ and $b \not\sqsubseteq a$
- totally[27] ordered set, to$S$: poSet only with relation pairs being comparable
  ex.: $R_{[\sqsubseteq]} :=$ "⊇" (as presented), $S := \{\{\}, \{0\}, \mathcal{B}\}$; $\forall a, b \in S$: $a \sqsubseteq b$ or $b \sqsubseteq a$
- chain: subset $M$ of a to$S$
  ex.: $S$ as defined above; $M := \{\{0\}, \mathcal{B}\} \subseteq$[28] $S$
- well-ordered set $S$: poSet$(S, \sqsubseteq)$, $\forall T_{[\neq\{\}]} \subseteq S\ \exists$ minimal element[29] $m$
  ex.: $S := \{1, 2\}$, $R_{[\sqsubseteq]} :=$ "≥". $T_1 := \{1\}$, $m_1 = 1$; $T_2 := \{2\}$, $T_3 := S$: $m_{2/3} = 2$
- *upper bound* (for $T_{[\subseteq S]}$) $b_u$: $\forall c \in T$: $c \sqsubseteq b_u$ [$\in$ poSet$(S, \sqsubseteq)$]
  ex.: $S := 2^{\mathcal{B}}$ (see above), $R_{[\sqsubseteq]} :=$ "⊆", $T := \{\{\}, \{0\}\}$; $b_u := \mathcal{B}$: $\{\}, \{0\} \sqsubseteq b_u$
- *least upper bound* $lub(T_S)_{[\in S]}$: $\forall b_u$: upper bound $lub(T_S) \sqsubseteq b_u$ [$\in$ poSet$(S, \sqsubseteq)$]
  ex.: $S$, $R_{[\sqsubseteq]}$, $T$ as just defined; $lub(T_S) = \{0\}$: $\{\}, \{0\} \sqsubseteq lub(T_S) \sqsubseteq_{\forall} b_u$
- *lower bound* (for $T_{[\subseteq S]}$) $b_l$: $\forall c \in T$: $b_l$ [$\in$ poSet$(S, \sqsubseteq)$] $\sqsubseteq c$
  ex.: $S$ and $R_{[\sqsubseteq]}$ as just defined, $T := \{\{0\}, \mathcal{B}\}$; $b_l := \{\}$: $b_l \sqsubseteq \{0\}, \mathcal{B}$

24 Exactly 1 of the two cases, not both together (however also not none).

25 Superset: in Section 2.2 (specified starting on page 14).

26 "Set of all subsets": see page 13, $\mathcal{B}$: cf. page 8, "{ }": page 12.

27 "Linearly": all elements can get ordered along a line and thereby compared according to an ordering.

28 Subset: specified in Section 2.2 (starting on p. 14).

29 This need not be unique; there just does not exist a "smaller" one related to the present "⊑"-relation, in the sense of being a/the "first" element.

- *greatest lower bound* $glb(T_S)_{[\in S]}$: $\forall b_l$: $b_l$ [$\in$ poSet$(S, \sqsubseteq)$] $\sqsubseteq$ lower bound $glb(T_S)$
  ex.: $S$, $R_{[\sqsubseteq]}$, $T$ as just defined; $glb(T_S) = \{0\}$: ${}_\forall b_l \sqsubseteq glb(T_S) \sqsubseteq \{0\}$, $\mathcal{B}$
- *lattice* $L$: poSet$(L, \sqsubseteq)$, $\forall (x,y) \in L^2$: $\exists\, lub(\{x,y\})$, $\exists\, glb(\{x,y\})$
  ex.: $R_{[\sqsubseteq]}$ := "$\subseteq$"; $V := \{1, \ldots, n\}$, $L := 2^V$ := set of all subsets of $V$.
  $lub(\{S_1, S_2\}) = S_1 \cup S_2$, $glb(\{S_1, S_2\}) = S_1 \cap S_2$.[30]
  In Figure 1.1, we regard the powerset-lattice for $n := 4$.

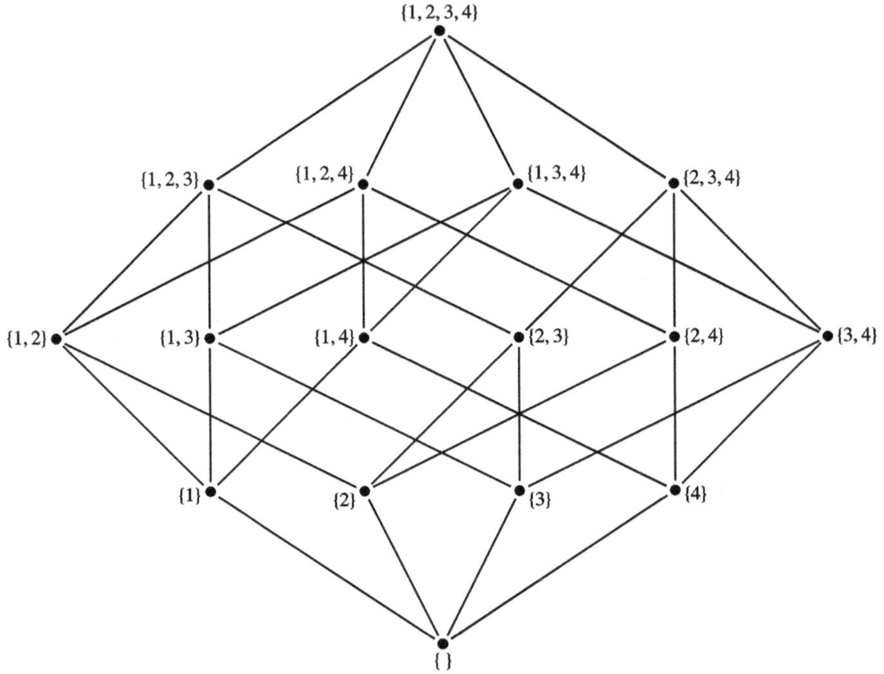


**Figure 1.1:** Powerset-lattice of a 4-elementary set with all $2^4$ (= 16) subsets (incl. $V$ ☺).

Each line represents a relation between two sets: from a certain level to the next higher one, the feature "proper subset," from a level to the next lower one "proper superset"; further sub/superset relations result via transitivity—inherent in the present preorder feature (cf. p. 9). You can yield this "transitive closure" of a relation w. r. t. a set $M$ regarding this picture starting from $M$ by just traversing the levels along the lines in constant direction on all possible paths. Even the aforementioned *lub* and *glb* can easily be visualized in this picture: Starting from two given sets $S_1$ and $S_2$, we visit the first "common" set; in order to build the *lub* to go upwards to the smallest set, which contains all elements from the two "start" sets,

30 "∪" ("set union") and "∩" ("intersection") get specified in Section 2.2 (starting on p. 15).

to build the *glb* going downwards to the largest set whose elements are contained in both start sets—as illustrated in the following three examples:

1. $S_1 := \{1\}, S_2 := \{3,4\}$; $lub(\{S_1, S_2\}) = \{1,3,4\}$, $glb(\{S_1, S_2\}) = \{\,\}$:
   *lub*: starting from $\{1\}$, we could visit $\{1,3\}$ or $\{1,4\}$ to reach $\{1,3,4\}$,
   from $\{3,4\}$ in one step immediately to $\{1,3,4\}$;
   *glb*: starting from $\{1\}$, we immediately visit $\{\,\}$,
   from $\{3,4\}$ via $\{3\}$ or $\{4\}$ (not sooner) to $\{\,\}$.
2. $S_1 := \{1,2\}, S_2 := \{2,3\}$; $lub(\{S_1, S_2\}) = \{1,2,3\}$, $glb(\{S_1, S_2\}) = \{2\}$:
   *lub*: starting from $\{1,2\}$, we immediately visit $\{1,2,3\}$,
   from $\{2,3\}$ as well;
   *glb*: from $\{1,2\}$, we reach in one step $\{2\}$, from $\{2,3\}$ as well.
3. $S_1 := \{1\}, S_2 := \{1,2\}$; $lub(\{S_1, S_2\}) =_{[S_2 \supseteq S_1]} S_2$, $glb(\{S_1, S_2\}) =_{[S_1 \subseteq S_2]} S_1$:
   *lub*: from $\{1\}$, we reach in a single step $\{1,2\}$, in $\{1,2\}$ we are already there;
   *glb*: in $\{1\}$, we are already there, from $\{1,2\}$ we reach (in a single step) $\{1\}$.
   In this final example, we insightfully see the use also of the other feature of the preorder (beside the transitivity, see p. 9), namely the reflexivity: a set is both ("improper") sub and superset to itself.

This all may come across very abstractly; however, by that the table is now neatly laid.

# 2 Set theory

Here, we lay the foundation for a sensible set theory, introduce the terms most important for us and list common laws. Afterwards, we define the order of magnitude of a set and illustrate similarities of and differences between finite and infinite sets. The ("higher-order") infinity presented at the end of this chapter even prepares the *uncomputability* in Theoretical Computer Science. Finally, we do not shy away from the *Generalized Continuum Hypothesis.*

## 2.1 Basics

A standard *set* $S$ is a collection of unique elements (no copies). Sometimes, however, you need the functionality of the multiple presence of elements; such a structure is a called *multiset* (copies allowed).[1] Then you are also interested in the number of occurrences of the respective elements: This is called the corresponding *multiplicity.*[2] The special set with exactly 1 element is called a *singleton.*

We now introduce a denotation for the "expressiveness" of a set $S$—its cardinal number, more crisply called *cardinality* $=: |S|$. In a finite embedding, it denotes the number of ("#") the elements, and in an infinite one, their so-called "order of magnitude." The smallest set, the empty set $\{\} =: \emptyset$, naturally has the smallest cardinality:

$$|\emptyset| = 0.$$

A set is called *countably infinite*, if there is a bijection with $\mathcal{N}$.

(At the end of this section, we see that this is only the first level of infinity.)

A set $S_c$ is *countable*, if it does not go beyond ($|S_c| \leq |\mathcal{N}|$):

- $0 \leq |S_c| \leq i_{[\in\mathcal{N}]} < |\mathcal{N}|$: $S_c$ finite;
- $0 \leq i_{[\in\mathcal{N}]} < |S_c| = |\mathcal{N}|$: $S_c$ infinite.

Now let us move on to something very fundamental in the area of sets and functions:

$$|A| = |B| \quad \Longleftrightarrow \quad \exists\texttt{bijection}\, f: A \to B.$$

It is clear in the finite world: If the number of elements in the sets is different, not every element from the larger set has an exclusive partner in the smaller one: no bijection.[3] However, if each set has the same number of elements, then every element can have a

---

1 In Subsection 5.4.1, we treat the # (so-called) permutations of elements both in their "unique" appearance as well as in the case where elements appear several times (starting on p. 71).

2 In a nonempty standard set for each element always 1.

3 This is like real life, which then has so-called "work-arounds." ◡̈.

https://doi.org/10.1515/9783111206899-003

unique partner; there might exist at least 1 bijection.[4] (Warning: Given a function $f$ just with a common cardinality of both its domain $D$ and codomain $C$, need not be a bijection: In such a negative case, $f$ always is neither injective nor surjective—illustration: $D := \mathcal{B} =: C$ with $\forall d \in D : f(d) := 0 \in C$.)

Things become wilder in an infinite setting: Here, you could create a bijection in certain constellations, even if by a first naive view a set seems to have fewer elements than the other, as it might be the case with a proper (here infinitely large) subset (of its corresponding superset).[5] An example would be the following.

Let $E_{[\subset \mathcal{N}]}$ := set of *e*ven natural numbers; then the following holds:

$$|E| = |\mathcal{N}|.$$

The typical bijective *function* $f$ is the following: $f : E \to \mathcal{N}$:

$$\begin{aligned} f(0) &:= 0 \\ f(2) &:= 1 \\ f(4) &:= 2 \\ &\vdots \\ f(e) &:= e/2. \end{aligned}$$

Let us inspect these corresponding two features (needed), *injectivity* and *surjectivity*: Different *e*ven numbers get injectively assigned different *n*atural numbers. No $n$ gets forgotten; each *n*atural number is surjectively reached by an *e*ven number as the function value. We confirm that both sets (although proper sub- and supersets of each other) have the same common cardinality/order of magnitude.[6] (Similarly, it would be possible to bijectively match the set of fractions $\mathcal{Q}$ with $\mathcal{N}$ showing $|\mathcal{Q}| = |\mathcal{N}|$ [uninteresting here].)

The comparison of the cardinalities of two infinitely large sets might also turn out quite differently and can be inspected in this passage:

A very important set is the *set of all subsets*[7] of a (base) set $S$—"power set" of $S$:

$$\mathcal{P}(S) := \{s \mid s \subseteq S\}$$

—sometimes denoted by $2^S$, due to this reason: Given $|S|$; then for finite sets the following statement holds:[8]

---

4 That variety ☺ could be offered is explained in Chapter 5 on the pages 71 and 83.

5 "Proper" (⊂) and "improper" (⊆) subsets (+ supersets) are specified in the following Section 2.2 .

6 Therefore, ("⊂", "⊃"), in the infinite embedding, we do not speak about # (elements).

7 See the aforementioned footnote 5, by the characteristic "improper subset" (⊆).

8 See the Subsections 4.1.1 (p. 40, "5," final part) and 5.4.2 (p. 76, specialized binomial theorem).

$$|\mathcal{P}(S)| = |2^S| = 2^{|S|} \quad [> 0].$$

The following further fact, which is true both for finite as well as infinite sets, has a fundamental meaning for our end (◡̈ of this chapter):

$$|\mathcal{P}(S)| > |S| \quad [\geq 0].$$

This is easy to see in a finite setting: Every present element $e$ of $S$ forms the singleton $\{e\} \in \mathcal{P}(S)$; additionally, we still have at least the empty set (which does not contain any element) $\{\}$ $[\in \mathcal{P}(S)]$ that always acts as subset of whatever set $S$ (even of its own)—therefore $\mathcal{P}(S)$ contains at least 1 more element than $S$.

In an infinite setting, the "$>$" means that there must exist a higher-order infinity than the one of the well-known set ($S :=$) $\mathcal{N}$ of the *n*atural numbers.[9] This is precisely one of the sources of the *uncomputability* in Theoretical Computer Science, which is easier to understand once we lay, as here in the current chapter, the foundations early. In contrast to just countably infinitely large sets (like $\mathcal{N}$ and the set $E$ priorly defined), which can be mapped bijectively onto each other, the introduced *p*ower set $\mathcal{P}(\mathcal{N})$ is a nice example of a so-called "uncountably" infinitely large set: The just countably infinite natural numbers are too few to bijectively assign each element of the set of all subsets of $\mathcal{N}$ an element from $\mathcal{N}$; these two sets have different cardinalities. We expand this discussion later here in Section 2.5 (on p. 25).

## 2.2 Terms

We start with the concept (already mentioned) of a *subset*:

$$A \subseteq B : \forall x \in A \quad \Longrightarrow \quad x \in B.$$

With regard to their cardinalities the following fact is obvious: $|A| \leq |B|$.

$A$ and $B$ might even be identical—the reason why we speak about an *improper* subset.

Once this special case is excluded, we name this set inclusion *proper*:

$$A \subset B : A \subseteq B \quad \text{and} \quad A \neq B \quad \Longleftrightarrow \quad A \subseteq B \quad \text{and} \quad \exists x \in B, \notin A.$$

For finite sets, the proper subset has less elements then the other one: $|A| < |B|$.

However, infinite sets might have the same common cardinality; see Section 2.1 (p. 13).

9 Even more, there are an infinite number of infinities—as illustrated at the end of the chapter.

The opposing relationship is called *superset*:

$$A \supseteq B : \forall x \in B \quad \Longrightarrow \quad x \in A.$$

With regard to their cardinalities, the following fact is obvious: $|A| \geq |B|$.

*A* and *B* might even be identical, the reason why we speak about an *improper* superset.

Once this special case is excluded, we name the superset *proper*:

$$A \supset B : A \supseteq B \quad \text{and} \quad A \neq B \quad \Longleftrightarrow \quad A \supseteq B \quad \text{and} \quad \exists x \in A, \notin B.$$

For finite sets, the proper superset obviously has more elements then the proper subset: $|A| > |B|$. Infinite sets might have the same common cardinality, as mentioned above.

Let us now treat *set equality*:

$$A = B \quad \Longleftrightarrow \quad A \subseteq B \quad \text{and} \quad A \supseteq B \quad \Longleftrightarrow \quad A \subseteq B \quad \text{and} \quad B \subseteq A.$$

Therefore, the following holds $\forall x : x \in A \Longleftrightarrow$ / <u>iff</u>[10] $x \in B$.

Now we describe the *intersection*:

$$A \cap B := \{x \mid x \in A \ \text{ and } \ x \in B\}.$$

Similarly, we describe the *union*:

$$A \cup B := \{x \mid x \in A \ \text{ or } \ x \in B\}.$$

For finite sets, its cardinality is as follows:

$$|A \cup B| = |A| + |B| - |A \cap B|.$$

The difference corrects the counting of the elements in the intersection; otherwise, those ones would get included twice. (The aforementioned "or," of course, includes the case of being there twice.) Should *A* and *B* be intersection-free ("disjoint") and, therefore, do not have any common element, we yield (in this special case):

$$A \cap B = \emptyset : |A \cup B| = |A| + |B|.$$

Let us now expand our discussion and regard the so-called "*u*niverse" as the set of all possible elements. Given a set $A$ ($\subseteq U$), we are now interested in all the elements from $U$ without (=: \) $A$, i. e., which do not appear in $A$; we call this set its *complement*:

---

**10** <u>i</u>f and only i<u>f</u> ["if": $x \in A \Leftarrow x \in B$; "only if": $x \notin B \Rightarrow x \notin A$, which means: $x \in A \Rightarrow x \in B$].

$$\bar{A} := U \setminus A := \{x \in U_{[\supseteq A]} \mid x \notin A\} =: A^c.$$

(We come back to the "\" operator later on again.)

Due to the fact that $A$ and $A^c$ "complements" each other to $U$ ($A \cup A^c = U$) and do not have any common elements (according to the definition), for finite sets we get the following cardinalities:

$$|A| + |A^c| = |U| \quad \Longleftrightarrow \quad |A^c| = |U| - |A|.$$

Now we come back to the "\" already used, the sign for the *set difference*:

$$A \setminus B := \{x \in A \mid x \notin B\} = A \cap B^c.$$

For finite sets, the following holds:

$$|A \setminus B| = |A| - |A \cap B|.$$

In the case that the set in front of the "\" is a superset of the one behind this operator (like the universe $\supseteq \ldots$), in this special case, we get

$$A \supseteq B : |A \setminus B_{[\subseteq A]}| = |A| - |B| \quad (=_{[A=:U_{(\supseteq B)}]} |B^c|),$$

due to the fact that $B$ is identical to the intersection with its superset $A_{[\supseteq B]}$.

A very interesting concept is the *symmetrical difference*:

$$A \oplus B := (A \cup B) \setminus (A \cap B).$$

Here, we collect those elements, which get considered in each of the sets exclusively, i. e., not belonging to both ($\notin$ "$\cap$").

Due to the fact that when counting the # elements in the union the elements in the intersection get already subtracted $1\times$ (such that they get not considered twice), we have to subtract this number again, in order to completely delete them (such that they really disappear). For finite sets, the cardinality is therefore as follows:

$$|A \oplus B| = |A| + |B| - 2 \cdot |A \cap B|.$$

Let us now focus on different sets without any intersection; such sets are called *disjoint*:

$$\nexists x \in A \cap B_{[\neq A]} \quad [= \emptyset].$$

A so-called *set family* is a collection of sets, which can be considered being in a certain connection. Is it possible to entirely "partition" a given set $S$ completely as a collection $P_S$ of $p$ nonempty subsets $A_i$, which all are disjoint to each other (having not a single

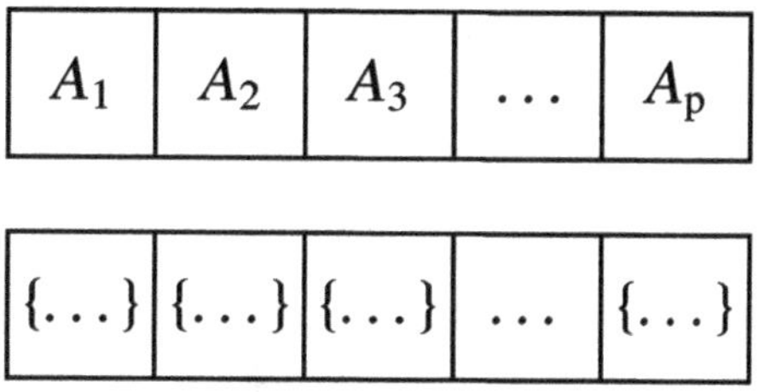


**Figure 2.1:** *p*-ary partition.

common element in whatever intersection theoretically possible); however, producing by the ∪ operator the given set $S$, we have a *p-ary partition*; see Figure 2.1:

$$P_S := \{A_1, A_2, A_3, \ldots, A_p\}, \quad p := |P_S|;$$

$$\forall_{[1\leq]} i \neq j_{[\leq p]} : A_i \cap A_j = \emptyset, \quad \bigcup_{i:=1}^{p} A_i = S.$$

We do not have to consider intersection elements; for finite sets the following holds:

$$|S| = \left|\bigcup_{i:=1}^{p} A_i\right| = \sum_{i:=1}^{p} |A_i|.$$

## 2.3 Laws

Now we introduce the 10 probably most well-known features of finite sets.

- Complement:[11] $A \cap A^c = \emptyset$; $A \cup A^c = U$.
- Double complement: $(A^c)^c = A$.
- Commutativity: $A \cap B = B \cap A$; $A \cup B = B \cup A$.
- Associativity: $(A \cap B) \cap C = A \cap (B \cap C)$; $(A \cup B) \cup C = A \cup (B \cup C)$.
- Dominance: $\emptyset \cap A = \emptyset$; $U \cup A = U$.
- Identity: $U \cap A = A$; $\emptyset \cup A = A$.
- Idempotence: $A \cap A = A$; $A \cup A = A$.
- Absorption: $A \cap (A \cup B) = A$; $A \cup (A \cap B) = A$.
- Distributivity: $A \cap (B \cup C) = (A \cap B) \cup (A \cap C)$; $A \cup (B \cap C) = (A \cup B) \cap (A \cup C)$. Figure 2.2 shows that a set intersection with a set union results in the same set when first producing two special intersections and then forming the set union of these two sets; Figure 2.3 illustrates that a set union with an intersection yields the identical set when first building two special unions and then producing the intersection of these two sets.

11 "completion": $A^c$ (mentioned above) completes $A$ to $U$.

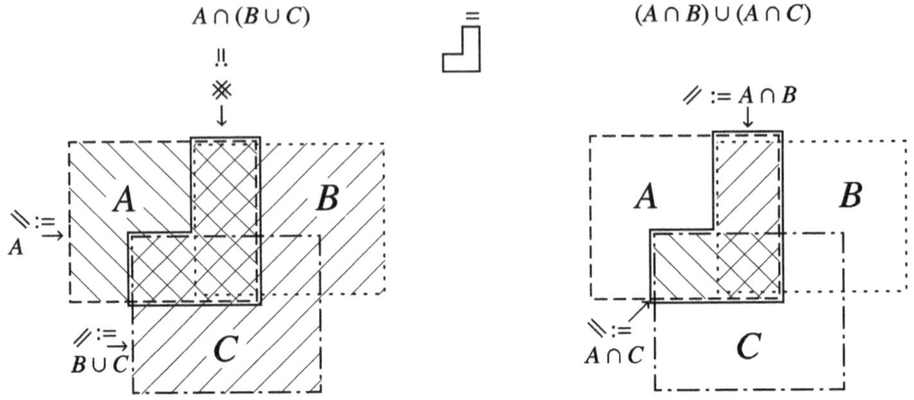


**Figure 2.2:** Set intersection with a union = Set union with special intersections.

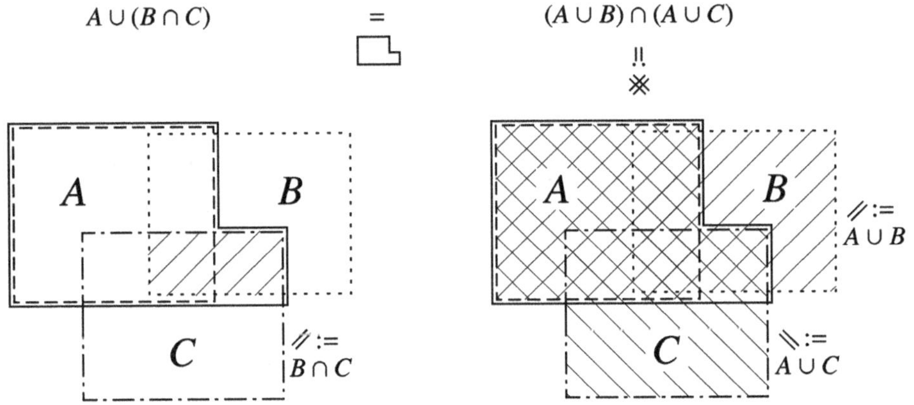


**Figure 2.3:** Set union with an intersection = Set intersection with special unions.

- *De Morgan* $(\bigcap_{i:=1}^{n} S_i)^c = \bigcup_{i:=1}^{n}(S_i^c)$; $(\bigcup_{i:=1}^{n} S_i)^c = \bigcap_{i:=1}^{n}(S_i^c)$.
  Figure 2.4 illustrates the equivalence of the complement of the intersection with the union of the individual complement sets, and Figure 2.5 the other way around, namely the equivalence of the complement of the union with the intersection of the individual complement sets.

## 2.4 Cardinality of finite sets

We enlighten the helpful method of counting elements via a *partition*.

- $U := A \ [\supset B_{\neq\{\}}], P_U := \{A \backslash B, B\}$
  Figure 2.6 shows the counting of elements of such a set difference.

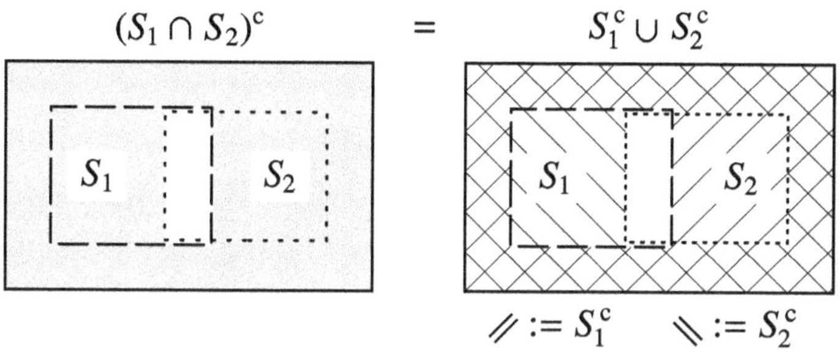


**Figure 2.4:** Intersection complement = Union of complements.

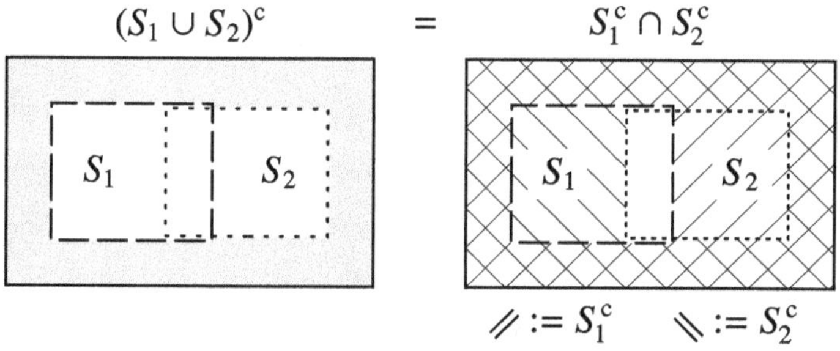


**Figure 2.5:** Union complement = Intersection with complements.

$$|A| = |A\backslash B| + |B| \quad \Longleftrightarrow \quad |A\backslash B| = |A| - |B|$$

- $U := A \cup B, P_U := \{A, B \setminus (A \cap B)\}$
  Figure 2.7 shows the counting for the union of disjoint parts.

$$|A \cup B| = |A| + \big|B \setminus (A \cap B)\big| =_{[B \supset A \cap B]} |A| + |B| - |A \cap B|$$

- $U := A \cup B, P_U := \{A \oplus B, A \cap B\}$
  Figure 2.8 shows the cardinality of the *XOR*; here, we produce it on 2 different ways.

$$\begin{aligned} |A \cup B| &= |A \oplus B| + |A \cap B| \quad \Longleftrightarrow \\ |A \oplus B| &= |A \cup B| - |A \cap B| \\ &= |A| + |B| - |A \cap B| - |A \cap B| \\ &= |A| + |B| - 2 \cdot |A \cap B| \\ &= \big(|A| - |A \cap B|\big) + \big(|B| - |B \cap A|\big) \\ &= |A \setminus B| + |B \setminus A| \end{aligned}$$

- $U := A \cup B, A \subset B$ or $A \supset B$; $|U| = \max.\{|A|, |B|\}$:
  a) $A \subset B =: U, P_U := \{B\backslash A, A\}$

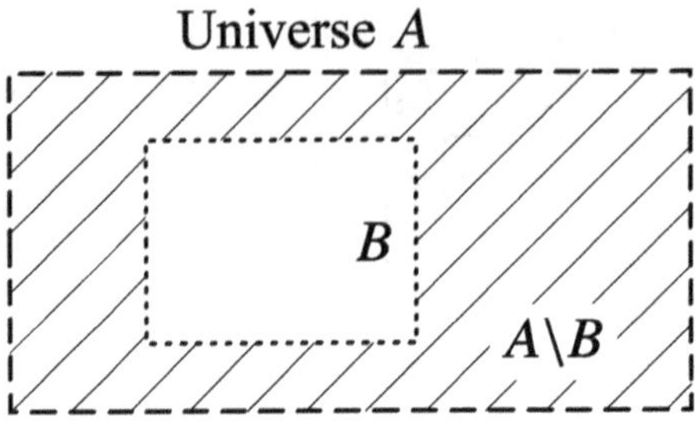


**Figure 2.6:** Cardinality of a set difference between a proper super- to a proper subset.

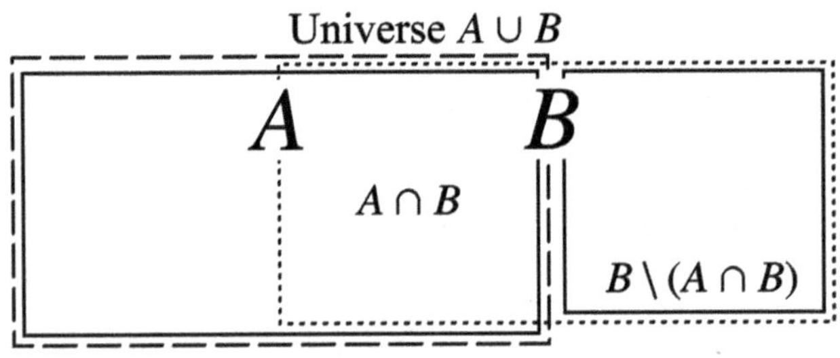


**Figure 2.7:** Union with disjoint sets (both parts of this partition).

$|U| = |B\backslash A| + |A| =_{[B\supset A]} |B| - |A| + |A| = |B| = \max.\{|A|, |B|\}$

b) $B \subset A =: U, P_U := \{A\backslash B, B\}$

$|U| = |A\backslash B| + |B| =_{[A\supset B]} |A| - |B| + |B| = |A| = \max.\{|A|, |B|\}$

– $U := A_1 \cup A_2 \cup A_3,\ P_U :=_{[\text{Fig. 2.9}]} \{//, \backslash\backslash, \|, =, \bullet\bullet, \square\square, \triangle\triangle\}$
$= \{A_1 \backslash (A_2 {\cup} A_3), A_2 \backslash (A_1 {\cup} A_3), A_3 \backslash (A_1 {\cup} A_2), (A_1 {\cap} A_2) \backslash A_3, (A_1 {\cap} A_3) \backslash A_2, (A_2 {\cap} A_3) \backslash A_1,$
$A_1 \cap A_2 \cap A_3\}$
Figure 2.9 prepares the formula for the cardinality of the union of three sets.

$$
\begin{aligned}
&|A_1 \cup A_2 \cup A_3| \\
&\quad = (|A_1| - |A_1 \cap (A_2 \cup A_3)|) + (|A_2| - |A_2 \cap (A_1 \cup A_3)|) + (|A_3| - |A_3 \cap (A_1 \cup A_2)|) \\
&\qquad + (|A_1 \cap A_2| - |(A_1 \cap A_2) \cap A_3|) + (|A_1 \cap A_3| - |(A_1 \cap A_3) \cap A_2|) \\
&\qquad + (|A_2 \cap A_3| - |(A_2 \cap A_3) \cap A_1|) + |A_1 \cap A_2 \cap A_3| \\
&\quad = |A_1| - |(A_1 \cap A_2) \cup (A_1 \cap A_3)| \\
&\qquad + |A_2| - |(A_2 \cap A_1) \cup (A_2 \cap A_3)| \\
&\qquad + |A_3| - |(A_3 \cap A_1) \cup (A_3 \cap A_2)| \\
&\qquad + |A_1 \cap A_2| - |A_1 \cap A_2 \cap A_3| \\
&\qquad + |A_1 \cap A_3| - |A_1 \cap A_2 \cap A_3| \\
&\qquad + |A_2 \cap A_3| - |A_1 \cap A_2 \cap A_3| \\
&\qquad + |A_1 \cap A_2 \cap A_3|
\end{aligned}
$$

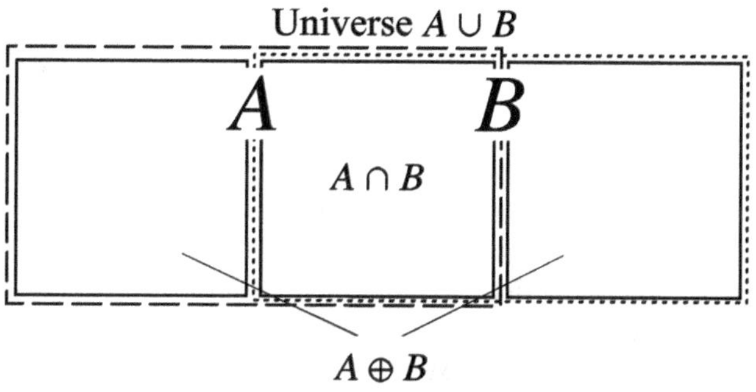


**Figure 2.8:** $|XOR|$.

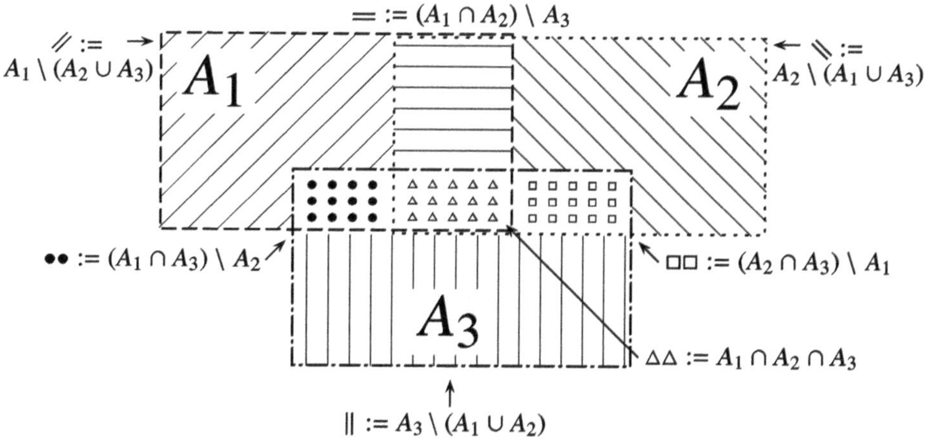


**Figure 2.9:** Multi-union.

$$
\begin{aligned}
&= |A_1| - (|A_1 \cap A_2| + |A_1 \cap A_3| - |A_1 \cap A_2 \cap A_3|) \\
&\quad + |A_2| - (|A_1 \cap A_2| + |A_2 \cap A_3| - |A_1 \cap A_2 \cap A_3|) \\
&\quad + |A_3| - (|A_1 \cap A_3| + |A_2 \cap A_3| - |A_1 \cap A_2 \cap A_3|) \\
&\quad + |A_1 \cap A_2| + |A_1 \cap A_3| + |A_2 \cap A_3| \\
&\quad + |A_1 \cap A_2 \cap A_3| - 3 \cdot |A_1 \cap A_2 \cap A_3| \\
&= |A_1| + |A_2| + |A_3| - 2 \cdot (|A_1 \cap A_2| + |A_1 \cap A_3| + |A_2 \cap A_3|) \\
&\quad + 1 \cdot (|A_1 \cap A_2| + |A_1 \cap A_3| + |A_2 \cap A_3|) \\
&\quad + 3 \cdot |A_1 \cap A_2 \cap A_3| \\
&\quad - 3 \cdot |A_1 \cap A_2 \cap A_3| \\
&\quad + |A_1 \cap A_2 \cap A_3| \\
&= |A_1| + |A_2| + |A_3| \\
&\quad - (|A_1 \cap A_2| + |A_1 \cap A_3| + |A_2 \cap A_3|) + |A_1 \cap A_2 \cap A_3|.
\end{aligned}
$$

What madness! Hopefully, the regularity just presented is not only optical; it would be great if we could incorporate this mechanism into a formula. Great—it is really the case: We start with the cardinalities of the individual sets, subtract the cardinality of all intersections of pairs[12] of sets and finally add the cardinality of the intersection among 3 sets.[13] This principle of "in/exclusion" (exactly in this order) is capable to get generalized to the computation of the # elements of the union of arbitrarily many sets;[14] it is subject of Section 5.2, which presents such a counting formula in its most general form immediately (without any intermediate calculation).

– $U := A_1 \cup A_2 \cup A_3$, $P_U := \{A_1 \backslash A_2, A_2 \backslash A_1, A_1 \cap A_2, A_3 \setminus (A_1 \cup A_2)\}$
*Hint*: First, create a sketch[15] and visualize this partition yourself:

$$\begin{aligned}
|A_1 \cup A_2 \cup A_3| &= |A_1 \backslash A_2| + |A_2 \backslash A_1| + |A_1 \cap A_2| + \big|A_3 \setminus (A_1 \cup A_2)\big| \\
&= |A_1| - |A_1 \cap A_2| \\
&\quad + |A_2| - |A_2 \cap A_1| \\
&\quad + |A_1 \cap A_2| \\
&\quad + |A_3| - \big|A_3 \cap (A_1 \cup A_2)\big| \\
&= |A_1| + |A_2| + |A_3| \\
&\quad - |A_1 \cap A_2| \\
&\quad - \big|(A_3 \cap A_1) \cup (A_3 \cap A_2)\big| \\
&= |A_1| + |A_2| + |A_3| \\
&\quad - |A_1 \cap A_2| \\
&\quad - (|A_1 \cap A_3| + |A_2 \cap A_3| - |A_1 \cap A_2 \cap A_3|) \\
&= |A_1| + |A_2| + |A_3| \\
&\quad - |A_1 \cap A_2| - |A_1 \cap A_3| - |A_2 \cap A_3| \\
&\quad + |A_1 \cap A_2 \cap A_3|.
\end{aligned}$$

## 2.5 Un/countability of infinite sets

Within this section, we now reach the highlight of the first part of this book. Let us start right now with something, which cannot exist for finite sets, namely that a proper superset should have the same cardinality as its proper subset.

---

12 $|\cup|$ of just 2 sets already subtracts the $|\cap|$ as well.

13 From 4 sets on, we would have to exclude all intersection elements stemming from the intersection of all 4 sets ....

14 Whether disjoint or not—as nicely illustrated above.

15 Please not in the book ⌣̈ —however, on using the only one, it could be funny ....

Each infinite set has an infinitely large proper subset with the identical cardinality:

$$S\ \texttt{infinite} \iff \exists T \subset S \quad \texttt{with} \quad |T| = |S|.$$

Let us for the moment consider the situation in this way: the infinitely large subset $T$ displayed above has a proper superset $S$ of identical cardinality, which sounds like having reached a certain end, regarding the order of magnitude of infinite sets. However, it goes beyond; life goes on ◡̈:

We construct the infinitely many discrete numbers set-theoretically and thereby consider each number as a "corresponding" set, creating the *ordinal* numbers ("ordinals"):

$$\begin{aligned}
0 &:= \emptyset \quad [= \{\,\}];\ 0^{\text{th}}\ \texttt{(finite) ordinal} =: \omega_0 \mathrel{\hat{=}} |\{\,\}| \\
\alpha + 1 =: \alpha^+ &\quad [\mathrel{\hat{=}} s(\alpha) := \texttt{successor}(\alpha)\,] := \alpha \cup \{\alpha\};\ \alpha\ \texttt{ordinal} \\
1 = 0 + 1 &:= \emptyset \cup \{\emptyset\} = \{\emptyset\} = \{0\} \mathrel{\hat{=}} |\{0\}| \\
2 = 1 + 1 &:= \{\emptyset\} \cup \{\{\emptyset\}\} = \{\emptyset, \{\emptyset\}\} = \{0, 1\} \mathrel{\hat{=}} |\{0, 1\}| \\
3 = 2 + 1 &:= \{\emptyset, \{\emptyset\}, \{\emptyset, \{\emptyset\}\}\} = \{0, 1, 2\} \mathrel{\hat{=}} |\{0, 1, 2\}| \\
&\ \vdots \\
\omega = \{0, 1, 2, \ldots\} =: \mathcal{N};&\quad 1^{\text{st}}\ \underline{\texttt{in}}\texttt{finite ordinal} =: \omega_1 \quad [\mathrel{\hat{=}} |\mathcal{N}|];
\end{aligned}$$

generally: $\alpha \mathrel{\hat{=}} |\alpha|$.

We see: $\alpha_1 < \alpha_2 \iff \alpha_1 \in \alpha_2 \overset{\text{here}}{\iff} \alpha_1 \subset \alpha_2$.

[Is this an exit from the set paradox of Bertrand Arthur William Russell? Let us try to artificially produce the equivalence just mentioned:

$$\alpha \not< \alpha \iff \alpha \notin \alpha \quad ◡̈.]$$

Now we introduce the *limit ordinal*—*lol* ◡̈; it does not have an immediate predecessor:

$$\beta\ \textit{lol} \iff \nexists \alpha \quad \texttt{with} \quad \beta = s(\alpha).$$

So far, we already know two of them: $\omega_0$ and $\omega_{\underline{1}}$, the finite and the first infinite lol. We proceed further on:

$$\begin{aligned}
\omega_1 + 1 &:= \omega_1 \cup \{\omega_1\} = \{\omega_1, \{\omega_1\}\} \\
&\ \vdots \\
\omega_1 + k &= \{0, 1, 2, 3, \ldots, \omega_1, \omega_1 + 1, \ldots, \omega_1 + k - 1\} \\
\omega_2 &= \omega_1 \cup \{\omega_1 + n \mid n \in \omega\} \quad 2^{\text{nd}}\ \texttt{infinite}\ \textit{lol}
\end{aligned}$$

$$\vdots$$

$$\omega_i = \omega_{i-1} \cup \{\omega_{i-1} + n \mid n \in \omega\} \quad i^{\text{th}} \texttt{ infinite } \textit{lol}$$

$$\vdots$$

(Ordinals form the basis for the *transfinite induction*, a proof[16] mechanism for the fixpoint semantics in PROLOG, a programming language with a high reputation especially in Europe and Japan in the area of *Intellectics/Artificial Intelligence*.)

In this way, we can easily produce infinitely many infinity levels. We could also create this scenario via consecutive constructions of the set of all subsets of the corresponding base set $\mathcal{N}$, as we see later on (p. 25). Once the cardinality of a set $S$ is larger than $|\mathcal{N}|$, then $S$ is called *uncountable*.

Rather interesting is the uncommutativity in the infinite world; already the addition of two ordinals, out of which at least one number is infinitely large, is not commutative. When riding through infinity, the order in which you change horses is important. ⌣̈ An example is the following:

$$\begin{aligned} 1+\omega: &\quad 1 < \omega \quad \Longrightarrow \quad 1 \subset \omega \quad \Longrightarrow \quad 1 \cup \omega = \omega; \\ \omega+1: &\quad 1 < \omega < \omega + 1 := \omega \cup \{\omega\} = \{\omega, \{\omega\}\} \neq \omega. \end{aligned}$$

This now holds: $1 + \omega \neq \omega + 1$.

Already thinking about a bijection between the set $\mathcal{Q}$ of fractions and the set $\mathcal{N}$ of natural numbers, we must be very careful during the traversal of the matrix[17] consisting of rows for the numerators ($\in \mathcal{Z}$) and columns for the denominators ($\in \mathcal{N}^+$) for the construction of the fractions: In the case we would constantly remain in one of the infinitely large rows (or columns), we would never reach all fractions; we must traverse it smartly, in this case in a diagonal manner (not part of our discussion here), and thereby actually yield a bijection, which further proves the possibility of having a common cardinality of a superset to its infinitely large proper subset.[18]

Let us finally develop the important feature regarding different cardinalities (orders of magnitude) of a very special kind of infinitely large sets "in the large"; as described in Section 2.1 (starting on p. 14): Each arbitrarily large set has a lower order of magnitude than its $\mathcal{P}$ower-set:

$$\begin{aligned} &S_1 := \omega_1 \\ &\omega_1 \mathrel{\hat{=}} |S_1| < |\mathcal{P}(S_1)| \\ &S_2 := \mathcal{P}(S_1) \end{aligned}$$

---

16 Via case analysis regarding successor and limit ordinal.

17 In both dimensions (countably) infinitely large.

18 As already mentioned on page 13, which is not always the case: $|\mathcal{R}| > |\mathcal{N}|$.

$$\omega_2 \mathrel{\hat{=}} |S_2| < |\mathcal{P}(S_2)|$$
$$S_3 := \mathcal{P}(S_2)$$
$$\omega_3 \mathrel{\hat{=}} |S_3| < |\mathcal{P}(S_3)|$$
$$\vdots$$
$$S_i := \mathcal{P}(S_{i-1}) \quad [:= \{s \mid s \subseteq S_{i-1}\}], \quad i > 1$$
$$\omega_i \mathrel{\hat{=}} |S_i| < |\mathcal{P}(S_i)|, \quad i \geq 0$$
$$\vdots$$

We remember that the $\mathcal{P}$ower set represents the set of all its subsets. Does this construction principle gets applied to an infinitely large set $S_{i-1}$, its "power"/expressiveness becomes visible impressively; for $i > 1$ $\mathcal{P}(S_{i-1})$, in comparison to $S_{i-1}$, enters exactly one higher level of infinity—by which we have a further way for the construction of $\omega_i$ $[\mathrel{\hat{=}} |\mathcal{P}(\omega_{i-1})|]$. We assume the Generalized Continuum Hypothesis[19]: $|\mathcal{N}| < |\mathcal{R}| \mathrel{\hat{=}} \omega_2 \mathrel{\hat{=}} |\mathcal{P}(\mathcal{N})|$.

$$\omega_{i-1} <_{[i>0]} \omega_{i-1} \cup \{\omega_{i-1} + n \mid n \in \omega\} = \omega_i \mathrel{\hat{=}}_{[i>1]} |\mathcal{P}(\omega_{i-1})|;$$

displayed in another way:

$$\omega_{i+1} \mathrel{\hat{=}}_{[i\geq 1]} |\mathcal{P}^i(\mathcal{N})|,$$

the cardinality of the subsequent execution of the $\mathcal{P}$owerset construction exactly $i$ times, starting from the smallest infinite set $\mathcal{N}$ (which represents the first level of infinity).

The importance of the fact of having sets with different orders of magnitude (regarding their infinite cardinalities), mentioned at the beginning of this chapter (Section 2.1, p. 14), in relation to Theoretical Informatics[20], gets now enlightened more carefully: Let $\Sigma^*$ be the infinite set of all possible "words" (strings) on a nonempty finite alphabet $\Sigma$, then $\mathcal{P}(\Sigma^*)$ represents the even larger set of all possible subsets, in this case called "languages". Once we now consider the "words" as "algorithms" and "languages" as "problems", we could understand that there are more problem instances than solving procedures. Hence, there exist problems which cannot be solved by algorithms, because a surjection from the set of algorithms in the set of languages is impossible[21]: The too "few" (just countably infinitely many) algorithms, due to the definition of a function, cannot cover all (uncountably many) languages.[22] We could bijectively assign each element from $\Sigma^*$ (reflecting a "word" [representing an algorithm]) an element from the

19 Just "the" (ordinary) continuum hypothesis Cantor-*CH*.

20 *Formal Languages* and *Computability Theory* (referring to "uncomputability"/"undecidability").

21 $\Longrightarrow \nexists$ bijection $\Sigma^* \to \mathcal{P}(\Sigma^*) \Longleftrightarrow |\Sigma^*| \neq_{[<]} |\mathcal{P}(\Sigma^*)|$.

22 In my lecture on *Formal Foundations* (of Computer Science), I even offer further considerations.

set $\mathcal{N}$atural numbers. $\mathcal{P}(\mathcal{N})$, at least regarding its cardinality, then may correspond to $\mathcal{P}(\Sigma^*)$, the set of problems, which—as depicted—has a higher cardinality than the set of algorithms. Hereby, we have a solid base for the *uncomputability* ⍨ in Theoretical Computer Science.

# 3 *Boole*an algebra

In this chapter, we introduce some basic terms, useful for propositional logic. Afterwards, we define the construction of composed propositions via truth tables. Finally, we present some laws (especially "equivalences").

## 3.1 Terms

Let $\mathcal{B} := \{0, 1\}$ be the *Boole*an set of the two values 0 and 1, which get interpreted as the two "truth" values *false* ($f$) and *true* ($t$), respectively. These *Boole*an constants are named as "atoms" or "atomic formulae".

A propositional variable represents a so-called (0-ary) "predicate" (without any input parameter), which can get assigned the value $f$ or $t$. Via the use of "connectors", we build composed expressions ("well-formed formulae") in the following, for instance, via the *Boole*an variables $p$, $q$, $r$.

## 3.2 Truth tables

We introduce logical operators and demonstrate the huge number of assignments.

### 3.2.1 Basic patterns

- NOT negation $\neg$ (Figure 3.1)

| $p$ | $\neg p$ |
|---|---|
| 0 | 1 |
| 1 | 0 |

**Figure 3.1:** *NOT*.

- AND conjunction $\wedge$ (Figure 3.2)

| $p$ | $q$ | $p \wedge q$ |
|---|---|---|
| 0 | 0 | 0 |
| 0 | 1 | 0 |
| 1 | 0 | 0 |
| 1 | 1 | 1 |

**Figure 3.2:** *AND*.

https://doi.org/10.1515/9783111206899-004

The connector $\wedge$ here is situated in the middle of two *Boole*an variables and the reason why we speak of infix notation. Especially for longer uniform expressions, however, the prefix notation is recommended, where one places the connector sign in front of the variables:

$$p \wedge q =: \wedge(p, q);$$

$$p_1 \wedge p_2 \wedge p_3 \wedge \cdots \wedge p_n = \wedge(p_1, p_2, p_3, \ldots, p_n) =: \bigwedge_{i:=1}^{n} p_i.$$

– NAND ($\neg$ AND) Sheffer stroke | (Figure 3.3)

| $p$ | $q$ | $\neg(p \wedge q)$ |
|---|---|---|
| 0 | 0 | 1 |
| 0 | 1 | 1 |
| 1 | 0 | 1 |
| 1 | 1 | 0 |

**Figure 3.3:** *NAND.*

– OR (inclusive-OR) disjunction $\vee$ (Figure 3.4)

| $p$ | $q$ | $p \vee q$ |
|---|---|---|
| 0 | 0 | 0 |
| 0 | 1 | 1 |
| 1 | 0 | 1 |
| 1 | 1 | 1 |

**Figure 3.4:** *OR.*

$$p \vee q =: \vee(p, q);$$

$$p_1 \vee p_2 \vee p_3 \vee \cdots \vee p_n = \vee(p_1, p_2, p_3, \ldots, p_n) =: \bigvee_{i:=1}^{n} p_i.$$

– NOR ($\neg$ OR) Peirce arrow $\downarrow$ (Figure 3.5)

| $p$ | $q$ | $\neg(p \vee q)$ |
|---|---|---|
| 0 | 0 | 1 |
| 0 | 1 | 0 |
| 1 | 0 | 0 |
| 1 | 1 | 0 |

**Figure 3.5:** *NOR.*

– XOR exclusive-OR ⊕ (Figure 3.6)

| $p$ | $q$ | $p \oplus q$ |
|---|---|---|
| 0 | 0 | 0 |
| 0 | 1 | 1 |
| 1 | 0 | 1 |
| 1 | 1 | 0 |

**Figure 3.6:** *XOR.*

– implication conditional → (Figure 3.7)

| $p$ | $q$ | $p \to q$ |
|---|---|---|
| 0 | 0 | 1 |
| 0 | 1 | 1 |
| 1 | 0 | 0 |
| 1 | 1 | 1 |

**Figure 3.7:** Implication.

Several ways exist referring to this expression; the most common ones are:
- if $p$ then $q$
- $p$ implies $q$
- $q$ (always) if $p$
- $q$ follows from $p$
- $p$ sufficient for $q$
- $q$ necessary for $p$
- $p$ only if $q$

The part in front of the → is called "antecedent", the one at the end "consequent". Furthermore, there exist interesting logical equivalences, which might be helpful in proofs; additionally, we learn corresponding technical terms:

$$p \to q \iff \neg p \vee q \iff \neg q \to \neg p.$$

The final form is called "contrapositive" and is not identical with this one:

$$q \to p \iff \neg p \to \neg q.$$

With respect to $p \to q$, the first expression is called "converse" and the final one "inverse".

– equivalence biconditional ↔ (Figure 3.8)
Also here, several wordings exist:
- $p$ and $q$ equivalent
- $p$ and $q$ imply each other

| *p* | *q* | $p \leftrightarrow q$ |
|---|---|---|
| 0 | 0 | 1 |
| 0 | 1 | 0 |
| 1 | 0 | 0 |
| 1 | 1 | 1 |

**Figure 3.8:** Equivalence.

- *p* if and only if ("iff") *q*
- *p* sufficient and necessary for *q*

$$p \leftrightarrow q \quad \Longleftrightarrow \quad (p \rightarrow q) \wedge (q \rightarrow p).$$

Finally,
- *CNF*
  The *C*onjunctive *N*ormal *F*orm is a propositional conjunction consisting of disjunctions of "literals" (negative/positive *bool*ean variables).
- *k-SAT*
  In *k-SAT*ISFIABILITY in each disjunction there exist up to *k* literals in a CNF, the latter one now considered as propositional satisfiability problem. Here, the challenge is whether there exist corresponding truth values to fulfill the given formula (s. t. it evaluates to "*true*" (or to determine that all exponentially many assignment variants produce "*false*").[1] Once we focus on $k \in \{2, 3\}$, we yield two prominent special cases; between them there exists a so-called "phase transition", related to the degree of difficulty of the corresponding problem solving: 2- and 3-*SAT* probably belong to different complexity classes[2] in Theoretical Computer Science.[3]

### 3.2.2 Assignments

Given are $n :=$ different *Bool*ean variables; then the following holds:
1. # different codings: $2^n$ [binary representations of the naturals 0 to $2^n - 1$];
2. # different functions: $2^{(2^n)}$ [# codings are now even in the exponent].

Figure 3.9 allows a first insight into both formulae. The proofs are located in Subsection 4.1.1 starting on page 40 as illustrated examples ("5" and "6") for induction with natural numbers.

1 A solution to this most famous problem is, at least in principle, always computable (unfortunately just impractically), because the search space is finite (and exponentially huge).

2 To solve the individual problem, in general.

3 The first case is linear and the second one seems to be exponential (see the following Subsection 3.2.2).

| p | q | r | $f_b$ |
|---|---|---|---|
| 0 | 0 | 0 | 0, 1 |
| 0 | 0 | 1 | 0, 1 |
| 0 | 1 | 0 | 0, 1 |
| 0 | 1 | 1 | 0, 1 |
| 1 | 0 | 0 | 0, 1 |
| 1 | 0 | 1 | 0, 1 |
| 1 | 1 | 0 | 0, 1 |
| 1 | 1 | 1 | 0, 1 |

**Figure 3.9:** # codings + virtual scheme # Boolean functions.

## 3.3 Laws

In the following, we learn the most well-known *Boole*an laws.

- contradiction: $p \wedge \neg p \Longleftrightarrow \mathit{false}$
- tautology: $p \vee \neg p \Longleftrightarrow \mathit{true}$
- double negation: $\neg\neg p \Longleftrightarrow p$
- commutativity: $p \wedge q \Longleftrightarrow q \wedge p$; $p \vee q \Longleftrightarrow q \vee p$
- associativity: $(p \wedge q) \wedge r \Longleftrightarrow p \wedge (q \wedge r)$; $(p \vee q) \vee r \Longleftrightarrow p \vee (q \vee r)$
- distributivity: $p \wedge (q \vee r) \Longleftrightarrow (p \wedge q) \vee (p \wedge r)$; $p \vee (q \wedge r) \Longleftrightarrow (p \vee q) \wedge (p \vee r)$
- dominance: $\mathit{false} \wedge p \Longleftrightarrow \mathit{false}$; $\mathit{true} \vee p \Longleftrightarrow \mathit{true}$
- identity: $\mathit{true} \wedge p \Longleftrightarrow p$; $\mathit{false} \vee p \Longleftrightarrow p$
- idempotence: $p \wedge p \Longleftrightarrow p$; $p \vee p \Longleftrightarrow p$
- absorption: $p \wedge (p \vee q) \Longleftrightarrow p$; $p \vee (p \wedge q) \Longleftrightarrow p$
- *De Morgan*: $\neg(\bigwedge_{i:=1}^{n} p_i) \Longleftrightarrow \bigvee_{i:=1}^{n}(\neg p_i)$; $\neg(\bigvee_{i:=1}^{n} p_i) \Longleftrightarrow \bigwedge_{i:=1}^{n}(\neg p_i)$
- exportation: $p \rightarrow (q \rightarrow r) \Longleftrightarrow (p \wedge q) \rightarrow r$

The first eleven laws are easy to grasp; the final equivalence could get proven gently:

$$p \rightarrow (q \rightarrow r) \quad \Longleftrightarrow \quad p \rightarrow (\neg q \vee r) \quad \Longleftrightarrow$$
$$\neg p \vee \neg q \vee r \quad \texttt{[=: left-hand side]} \Longleftrightarrow \texttt{[right-hand side :=]} \quad \neg(p \wedge q) \vee r \quad \Longleftrightarrow$$
$$(p \wedge q) \rightarrow r.$$

We remember the laws in the area of sets in Section 2.3;[4] they work in the *Boole*an Algebra correspondingly—a kind of "duality". Looking the other way around, we can recognize in the "exportation" rule a statement related to sets; after having formed the "*U*niverse", the aforementioned "left-" and "right-hand side" look as follows in the set version:

$$U := P \cup Q \cup R; \quad P^C \cup Q^C \cup R \quad \overset{\text{De Morgan}}{\Longleftrightarrow} \quad (P \cap Q)^C \cup R.$$

4 The two statements, which are clustered in the "complement" there, starting on page 17, have two different terms ("contradiction" and "tautology") here.

There exist further analogies between set theory and *Boole*an Algebra. To illustrate it a little bit ;) the # elements in the power-set of an $n$-elementary base set is identical to the number of rows in the truth table for $n$ *Boole*an variables: a “1” represents the fact that the element exists in the subset, a “0” the opposite. Hence, the empty set is represented by the coding “$(0, \ldots, 0)$”, the base set itself by “$(1, \ldots, 1)$”; similar representations exist for the mixed assignments. This is the reason why both the power-set as well as the truth table provide $2^n$ different entries.

# 4 Proof principles

This chapter introduces the three most common methods of mathematical proofs. We start with the basic technique *induction*, proceed with the *direct proof* and close with the *indirect proof*. (The *diagonalization*, which is important for Theoretical Computer Science, is placed in my German-written Informatics book, and might get incorporated in a potential second edition ☺.)

## 4.1 Induction

We discuss the *inductive* principle on the one hand on natural numbers related to both an original input $n$ and also to a logarithmic value (level number/search depth in a decision tree with uniform branching factor), and on the other hand, on strings ("words"), in order to proof typical informatics statements in formal language theory.[1] (I keep the interesting[2] *structural induction* for the aforementioned second edition[3]— to not overload the story in this introductory version.)

### 4.1.1 Natural numbers

The induction principle always follows the same scheme: At first, we show the validity for a very elementary *basis*[4] (=: $n_0$). Then we take the statement for some general case (e. g., $n - 1$) as valid *hypothesis* and finally produce in a constructive real-world *step* (e. g., from $n - 1$ to $n$) that thereby the hypothesis could get enhanced to the next larger structure (e. g., $n$), which exactly corresponds to the official statement.[5] (Notationally, the replacement of the predecessor structure by the formula of the hypothesis in the corresponding step gets signaled by "!".)

In doing so, we show that the structure of the proposition (which should get proven) is covered by the constructive problem-solving principle via the solution formula for

1 Different words get considered/sorted according to their length. The empty "word" $\varepsilon$ has the length 0, a word consisting of just 1 character (of a given alphabet) has the length 1, a word consisting of 2 characters the length 2, etc. Words of a common length get sorted lexicographically (typically in ascending order). In this way, we can provide to each arbitrary string a unique number—the reason why we, again, end up in the well-known set $\mathcal{N}$, and everything goes on as usual.

2 For Computer Science.

3 In the case this edition "runs" well ☺.

4 Ideally the smallest possible number (oftentimes 0 or 1, rarely 2 or 3, sometimes 4 or 5).

5 More formally: Let $A(n)$ be the proposition, which should get proven for any arbitrary $n$; in order to do that we first show $A(n_0)$ and then constructs, based on $A(n-1)$, with a problem-dependent successor step reaching $A(n)$. [Pupils school books might use the step from $n$ to $n+1$.]

https://doi.org/10.1515/9783111206899-005

infinitely many cases.[6]

The following examples illustrate this traditional proof technique:

1. Number of edges in the "complete graph"
   A (general) graph consists of a set $V$ of *v*ertices (*n*odes, $|V| =: n$) and a set $E$ of *e*dges ($|E| =: e$). In this special case now, we have an individual edge between each pair of nodes; there does not exist any orientation, i. e., we do not have directed connections ("arrows"), just undirected arcs.
   Let $n := |V|$, $e_n := |E|$; then the following holds:
   $$e_n = \frac{n \cdot (n-1)}{2}.$$
   Proof: Induction on $n$:
   (a) Basis: $n_0 := 1$
   Principle: $e_{1_P} = 0$ ($\nexists$ edge once we have only 1 node);
   Formula: $e_{1_F} = 1 \cdot (1-1)/2 = 0 = e_{1_P}$.
   The principle from the real world gets covered by the formula.
   (b) Hypothesis:
   $$e_{n-1} = \frac{(n-1) \cdot ((n-1)-1)}{2} \quad \left[= \frac{(n-1) \cdot (n-2)}{2}\right].$$
   (c) Step: $(n_0 \leq)\ n-1 \to n\ (> n_0)$
   Idea: The edges of the next smaller complete graph are still needed, and the node with the number $n$ gets connected to all $n-1$ nodes already there—by so many individual edges:
   $$\begin{aligned} e_{n_P} &= e_{n-1} + (n-1) \\ &\overset{!}{=} \frac{(n-1) \cdot (n-2)}{2} + \frac{2 \cdot (n-1)}{2} \\ &= \frac{(n-2+2) \cdot (n-1)}{2} \\ &= \frac{n \cdot (n-1)}{2} = e_{n_F}. \end{aligned}$$
   We yield the proof of the statement by the usage of a constructive step on top of the hypothesis and thereby demonstrate that the formula reflects the situation in the real world.
2. Number of nodes in the "decision tree"
   We have a tree with a uniform *b*ranching factor $b$ and deepest *l*evel $l$ (the depth, here reflecting the number of decision variables). In Figure 4.1, we see a binary

6 That such an approach makes sense is based on Peano axiom "5." (cf. page 2).

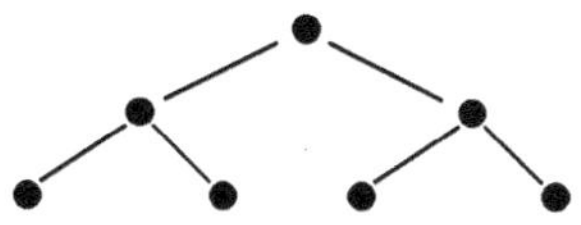

**Figure 4.1:** Decision tree.

tree ($b := 2$) with 3 levels, whose level numbers carry the labels 0 (root level), 1 (middle level, of course just here) and 2 (=: $l$, "*l*eaf level").
We can interpret it as follows: The root position represents our initial "point of view". Here, we branch for the decision regarding variable 1 according to "if-then-else" to the left to "then" and to the right to "else", at the next decision level regarding variable 2 to the left again to "then" and to the right again to "else", etc.; when we take all variables into consideration, we have on the (fina*l*) *l*eaf level $l$ exactly $2^l$ "leaves". Once we interpret "then" as *true* ("1") and "else" as *false* ("0"), we find at this level $l$ from right to left the binary encodings, starting to the right with the smallest number 0 further to the left reaching the largest number $2^l - 1$, by which we encode $2^l$ numbers. This corresponds to the number of rows of a *Boole*an truth table, which represents for $l$ variables all $2^l$ *false*/*true* assignment combinations.
Let us interest, for instance, to compute the memory demand in game-design, in the total number of nodes ("configurations") in the tree, when each level is completely occupied, i. e., having at every decision point the full degree of freedom.
Let $b_{[>1]}$ := *b*ranching factor (# different possibilities to proceed),

$l$ := search-depth (fina*l* decision level),

$s_b(l)$ := *s*um of all nodes across all levels (from 0 to $l$); the following holds

(*geometrical series*):

$$\sum_{i:=0}^{l} b^i = \frac{b^{(l+1)} - 1}{b - 1} = s_b(l).$$

Proof: Induction on $l$ [$= \log_b(b^l) \in \mathcal{N}$]:
(a) Basis: $l_0 := 0$

Principle: $s_b(0)_P = 1$ ($\exists!$ node: root);

Formula: $s_b(0)_F = (b^{(0+1)} - 1)/(b - 1) = (b - 1)/(b - 1) = 1 = s_b(0)_P$.

(b) Hypothesis:

$$s_b(l-1) = \frac{b^{((l-1)+1)} - 1}{b - 1} \quad \left[= \frac{b^l - 1}{b - 1}\right].$$

(c) Step: $(l_0 \leq)\ l - 1 \rightarrow l\ (> l_0)$

In addition to the sum so far, we now add on the final level $l$ still $b^l$ nodes:

$$\begin{aligned} s_b(l)_{\mathrm{P}} &= s_b(l-1) + b^l \\ &\stackrel{!}{=} \frac{b^l - 1}{b-1} + \frac{(b-1)\cdot b^l}{b-1} \\ &= \frac{b^l - 1 + b^{(l+1)} - b^l}{b-1} \\ &= \frac{b^{(l+1)} - 1}{b-1} = s_b(l)_{\mathrm{F}}. \end{aligned}$$

3. # levels completely occupied in the *Fibonacci*-tree
   Given is the following recursive construction of the *F*ibonacci numbers:

$$F_0 := 0, \quad F_1 := 1; \quad F_{n_{[>1]}} := F_{n-1} + F_{n-2}, \quad n \geq 2.$$

Let us compute $F_6$. This index 6 (our input size $n$) is not one of the basic indices 0 (in $F_0$) or 1 (in $F_1$); its $F$-value is not immediately present. We therefore replace the general index $n$ by the concrete input 6 and take the recursion as the computation procedure:

$$F_6 := F_{6-1} + F_{6-2}.$$

$F_5$ is not present either, the reason why we again call the *F*ib recursion, and so on. The resulting binary tree of recursion calls of the *F*ib-function is depicted in Figure 4.2:

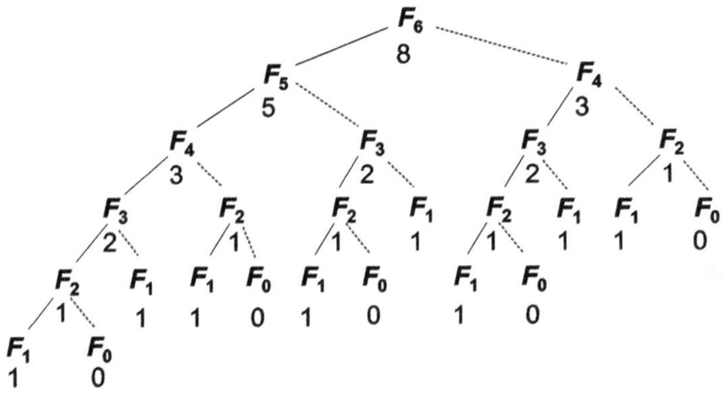


**Figure 4.2:** Fib-tree.

Let $c_n :=$ # completely occupied $F$-levels with input index $n$; the following holds:

$$c_n = 1 + \left\lfloor \frac{n}{2} \right\rfloor.$$

Proof: Induction on $n$:

(a) Basis:

i. $n_0 := 0$
   - $c_{0_P} = 1$ (top level: $F_0$),
   - $c_{0_F} = 1 + \lfloor \frac{0}{2} \rfloor = 1 = c_{0_P}$;

ii. $n_1 := 1$
   - $c_{1_P} = 1$ (top level: $F_1$),
   - $c_{1_F} = 1 + \lfloor \frac{1}{2} \rfloor = 1 = c_{1_P}$.

(b) Hypothesis:

$$c_{n-1} = 1 + \left\lfloor \frac{n-1}{2} \right\rfloor, \quad c_{n-2} = 1 + \left\lfloor \frac{n-2}{2} \right\rfloor.$$

(c) Step: $(n_0 \leq)\ n-2, n-1 \to n\ (> n_1 > n_0)$

Idea: We take the minimum of the two predecessor values, because we are interested just in the number of those levels completely occupied; due to the new input index $n$, we add "on top" 1 further top-level[7]—the reason why we add 1:

$$\begin{aligned}
c_{n_P} &= 1 + \mathtt{min}\{c_{n-1}, c_{n-2}\} \\
&\overset{!}{=} 1 + \mathtt{min}\left\{1 + \left\lfloor \frac{n-1}{2} \right\rfloor, \underline{1 + \left\lfloor \frac{n-2}{2} \right\rfloor}\right\} \\
&= 1 + \begin{cases} \frac{2+(n-2)}{2}; & \mathsf{even}(n) \\ \frac{2+((n-2)-1)}{2}; & \mathsf{odd}(n) \end{cases} \\
&= 1 + \begin{cases} \frac{n}{2}; & \mathsf{even}(n) \\ \frac{n-1}{2}; & \mathsf{odd}(n) \end{cases} \\
&= 1 + \left\lfloor \frac{n}{2} \right\rfloor = c_{n_F}.
\end{aligned}$$

According to the example ("2.") beforehand, the level number $c_n$ represents the exponent in the formula for the number of nodes in the main upper part of the tree—the reason why we easily recognize that already the number of levels completely occupied indicates that the number of *Fib* calls is exponential related to the input index: For the usage of the statement on page 35: $b := 2$, the highest *level* number $l$ is $c_n - 1 \implies l + 1 = c_n$, $s_b(l) = 2^{c_n} - 1 = 2^{(1+\lfloor n/2 \rfloor)} - 1$.

Obviously, there exists a linear algorithm, which—"bottom-up" (iteratively)—starting at the basic indices, proceeds step-by-step up to the input index $n$ and thereby, proportionally to the input size after $n$ steps, outputs $F_n$. However, we do not want to dig into algorithmics; we just explicate this induction principle.

---

7 With new root for the connection of the 2 existing (sub)trees with the indices $n-1$ and $n-2$.

Anyhow, there still exists a method, whose computation time might perhaps be even smaller once we could use fast exponentiation:[8]

4. Binet's closed *F*ib-formula
   Let $\phi := (1+\sqrt{5})/2$, $\psi := (1-\sqrt{5})/2$; then the following statement holds for the *F*ibonacci number with input index $n$:

$$f_n = \frac{\phi^n - \psi^n}{\sqrt{5}} \quad [\in_{\ddot{\smile}} \mathcal{N}].$$

Proof: Induction on $n$:

(a) Basis:
   i. $n_0 := 0$
      - $f_{0_P} = 0$ (first basic value),
      - $f_{0_F} = (\phi^0 - \psi^0)/\sqrt{5} = (1-1)/\sqrt{5} = 0 = f_{0_P}$;
   ii. $n_1 := 1$
      - $f_{1_P} = 1$ (second basic value),
      - $f_{1_F} = (\phi^1 - \psi^1)/\sqrt{5} = ((1+\sqrt{5})/2 - (1-\sqrt{5})/2)/\sqrt{5} = ((1+\sqrt{5}) - (1-\sqrt{5}))/(2\sqrt{5}) = (1-1+2\sqrt{5})/(2\sqrt{5}) = 1 = f_{1_P}$.

(b) Hypothesis:

$$f_{n-1} = \frac{\phi^{(n-1)} - \psi^{(n-1)}}{\sqrt{5}};$$

$$f_{n-2} = \frac{\phi^{(n-2)} - \psi^{(n-2)}}{\sqrt{5}}.$$

(c) Step: $(n_0 \leq)\, n-2, n-1 \to n\ (> n_1 > n_0)$

$$\begin{aligned}
f_{n_P} &= f_{n-1} + f_{n-2} \\
&\overset{!}{=} \frac{(\frac{1+\sqrt{5}}{2})^{(n-1)} - (\frac{1-\sqrt{5}}{2})^{(n-1)}}{\sqrt{5}} + \frac{(\frac{1+\sqrt{5}}{2})^{(n-2)} - (\frac{1-\sqrt{5}}{2})^{(n-2)}}{\sqrt{5}} \\
&= \frac{(\frac{1+\sqrt{5}}{2})^{(n-2)} \cdot (\frac{1+\sqrt{5}}{2} + 1) - (\frac{1-\sqrt{5}}{2})^{(n-2)} \cdot (\frac{1-\sqrt{5}}{2} + 1)}{\sqrt{5}} \\
&= \frac{(\frac{1+\sqrt{5}}{2})^{(n-2)} \cdot \frac{(1+\sqrt{5}+2)\cdot 2}{2\cdot 2} - (\frac{1-\sqrt{5}}{2})^{(n-2)} \cdot \frac{(1-\sqrt{5}+2)\cdot 2}{2\cdot 2}}{\sqrt{5}} \\
&= \frac{(\frac{1+\sqrt{5}}{2})^{(n-2)} \cdot \frac{1+2\sqrt{5}+5}{2^2} - (\frac{1-\sqrt{5}}{2})^{(n-2)} \cdot \frac{1-2\sqrt{5}+5}{2^2}}{\sqrt{5}} \\
&= \frac{(\frac{1+\sqrt{5}}{2})^{(n-2)} \cdot (\frac{1+\sqrt{5}}{2})^2 - (\frac{1-\sqrt{5}}{2})^{(n-2)} \cdot (\frac{1-\sqrt{5}}{2})^2}{\sqrt{5}} \\
&= \frac{\phi^n - \psi^n}{\sqrt{5}} = f_{n_F}.
\end{aligned}$$

8 We do not consider space.

[Digression: $\phi$—*Fib*—GCD]

– Binet's formula is not suitable in the area of discrete[9] numbers in a computer. We therefore try to compute $Fib_n$ step-by-step. Once we accept a slight deviation, we are able to base our computation of just 1 predecessor (without being forced to use 2 of them). At first, we recall $\phi$ just introduced:

$$\phi := \frac{1+\sqrt{5}}{2}.$$

Concerning the *F*ib number with index $i$ (incl. $n$, at the end), we observe:

$$\begin{aligned}
&\frac{f_{i-1}}{f_{i-2}} < \phi < \frac{f_i}{f_{i-1}}; \quad i := 2k+1,\ k \in \mathcal{N}^+;\\
&\lim_{n\to\infty} \frac{f_n}{f_{n-1}} = \phi \quad \left[\approx \frac{8}{5}\right],\\
&f_n \approx \phi \cdot f_{n-1}; \quad n \geq 6.
\end{aligned}$$

$\phi$ also appears in other areas; it is the "Golden Ratio". In architecture, it represents the division[10] of a front wall in 2 differently large parts such that the relation of the total length to the longer part is identical to the relation of the longer to the shorter part, as depicted in Figure 4.3:

$$\begin{aligned}
&t := l+s; \quad \frac{t}{l} = \frac{l}{s}.\\
&\phi := \frac{l}{s} = \frac{t}{l} \quad |\cdot ls \quad \Longleftrightarrow\\
&l^2 = ts \quad |-ts \quad \Longleftrightarrow\\
&l^2 - ts = 0 \quad \Longleftrightarrow\\
&l^2 - (l+s)s = 0 \quad \Longleftrightarrow\\
&l^2 - sl - s^2 = 0 \quad \Longleftrightarrow\\
&l = \frac{s}{2} \pm \sqrt{\left(-\frac{s}{2}\right)^2 - (-s^2)} \quad \Longleftrightarrow
\end{aligned}$$

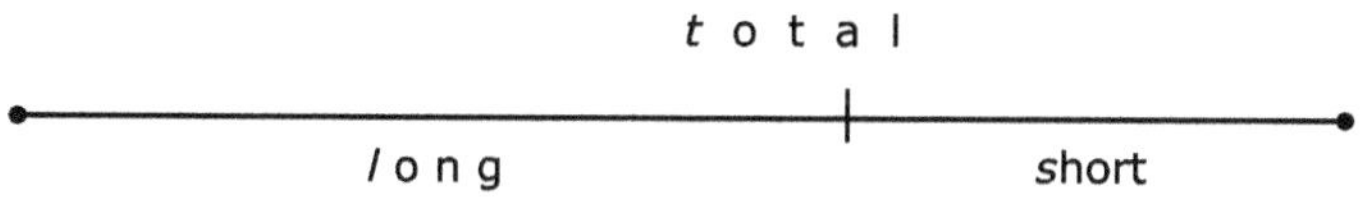


**Figure 4.3:** Golden ratio.

9 Due to the (square) root.

10 Considered to be "harmonious".

$$
\begin{aligned}
l &= \frac{s}{2} \pm \sqrt{\frac{s^2 + 4s^2}{4}} &&\Longleftrightarrow\\
l &= \frac{s}{2} \pm \sqrt{\left(\frac{s}{2}\right)^2 \cdot 5} &&\Longleftrightarrow\\
l &= \frac{s}{2} \cdot (1 \pm \sqrt{5}) \quad {\scriptstyle [>0]} &&\Longrightarrow\\
l &= \frac{1+\sqrt{5}}{2} \cdot s \quad | : s &&\Longleftrightarrow\\
\frac{l}{s} &= \frac{1+\sqrt{5}}{2} =: \phi := \frac{t}{l} && \ddot\smile .
\end{aligned}
$$

– Let GCD := Greatest Common Divisor of two numbers. The following holds:

$$Fib(\mathrm{GCD}(m,n)) =_{\text{proof here}}^{\text{without}} \mathrm{GCD}(Fib(m), Fib(n)).$$

Example:

$$
\begin{aligned}
m &:= 6, \quad n := 9;\\
Fib(\mathrm{GCD}(6,9)) &= Fib(\mathrm{GCD}(3\cdot 2,\ 3\cdot 3)) = Fib(3) = 2.\\
\mathrm{GCD}(Fib(6), Fib(9)) &= \mathrm{GCD}(8,34) = \mathrm{GCD}(2\cdot 4,\ 2\cdot 17) = 2.
\end{aligned}
$$

5. Number of rows in the *Bool*ean truth table
Let $r_n$ := rows of a table of values for $n$ *Bool*ean variables; the following holds:

$$r_n = 2^n.$$

Proof: Induction on $n$:
(a) Basis: $n_0 := 1$

$$
\begin{aligned}
r_{1_P} &= 2 \quad [= |\{\mathit{false}, \mathit{true}\}|];\\
r_{1_F} &= 2^1 = r_{1_P}.
\end{aligned}
$$

(b) Hypothesis:

$$r_{n-1} = 2^{(n-1)}.$$

(c) Step: $(n_0 \le)\ n-1 \to n\ (> n_0)$

$$r_{n_P} = 2 \cdot r_{n-1} \stackrel{!}{=} 2^1 \cdot 2^{(n-1)} = 2^n = r_{n_F}.$$

In the current example, the content of the $2^n$ rows represents the binary notation of the $2^n$ numbers 0 to $2^n - 1$. Having from the step before $2^{(n-1)}$ rows, the numbers so far still remain by adding a leading 0, and we get $2^{(n-1)}$ new larger numbers by

placing a leading 1 in front of, s. t. we end up with double the numbers (rows here) compared to the prior situation.
Furthermore, $2^n$ corresponds to the number of elements in the power-set, the set of all subsets of an $n$-elementary base set. The induction scheme used here (doubling the predecessor structure on the transition from $n-1$ to $n$) also appears in the construction of a new larger power-set: All the $2^{(n-1)}$ subsets already there remain, and to each of these subsets we further add the new element to each of the former subsets creating new ones—the reason why we have finally doubled the number of elements in the power-set, reflected by the constructive step mentioned beforehand [in (c)] "2·".

6. Number of *Boole*an functions
Let $d_n$ := # all different $n$-ary *Boole*an functions; the following holds:

$$d_n = 2^{(2^n)}.$$

Background: Similar to the situation before we take the total number of inputs into the exponent of the power to the base 2 (because each of the $2^n$ possible input-rows has a binary function-output).

Proof: Induction on $n$:
(a) Basis: $n_0 := 0$

$$d_{0_P} = 2 \quad [= |\{(\text{no } \underline{\text{input}}, \underline{f}alse_{\underline{\text{output}}}), (\text{no } \underline{\text{input}}, \underline{t}rue_{\underline{\text{output}}})\}|],$$
$$d_{0_F} = 2^{(2^0)} = 2^1 = d_{0_P};$$

or: $n_0' := 1$

$$d_{1_P} = 4 \quad [= |\{(f_{\text{in}}, f_{\text{out}}), (f_{\text{in}}, t_{\text{out}}), (t_{\text{in}}, f_{\text{out}}), (t_{\text{in}}, t_{\text{out}})\}|],$$
$$d_{1_F} = 2^{(2^1)} = 2^2 = d_{1_P}.$$

(b) Hypothesis:

$$d_{n-1} = 2^{(2^{(n-1)})}.$$

(c) Step: $(n_0 \leq)\ n-1 \to n\ (> n_0)$
With the next larger $n$, according to the example before, the # rows gets doubled, which is reflected in the current induction step:

$$\begin{aligned} d_{n_P} &= |\mathcal{B}|^{\#\,\text{rows}_n} \underset{5^{\text{th}}\text{ example}}{\overset{\text{step}}{=}} 2^{(2\cdot\#\,\text{rows}_{n-1})} \\ &= (2^{\#\,\text{rows}_{n-1}})^2 = (d_{n-1})^2 \overset{!}{=} (2^{(2^{(n-1)})})^2 \\ &= 2^{(2^{(n-1)}\cdot 2^1)} = 2^{(2^n)} = d_{n_F}. \end{aligned}$$

### 4.1.2 String length

Idea: The so-called "empty" word $\varepsilon$ has the length 0; once we already have a word and add a further character, we yield another word which has 1 more character (than the one before). In doing so, we could build all possible strings of arbitrary length. Although the alphabet itself just provides a finite number of characters (to construct words), infinitely many words are possible; each of them could get uniquely labeled by a natural number.

A potential induction proof therefore works along the length of whatever string.[11]

Let $a :=$ character ($\in$ alphabet), $\varepsilon$ the empty word; $u$, $v$, and $w$ are arbitrary words. The following holds:

$$\varepsilon\, w = w\, \varepsilon = w,$$
$$(u\, v)\, w = u\, (v\, w);$$
$$|w| := \texttt{length of the word}:$$
$$|\varepsilon| := 0,$$
$$|u\, v| = |u| + |v| \quad \Longrightarrow$$
$$|v\, a| = |v| + 1.$$

3 Remarks:

- Parentheses are not part of the alphabet; they just clarify the processing mode.
- The concatenation of words is not commutative;[12] hence, in general, the following holds: $u\, v \neq v\, u$ (without excluding special cases which might be commutative).
- In the case "*va*" above, we syntactically consider "*a*" being a string of length 1.

Application: string *r*eversal $\rho$: $\rho(c_1 c_2 \ldots c_n) = c_n \ldots c_2 c_1$, $c_{_{[1\leq]}i_{[\leq n]}}$ := character:

$$\rho(\varepsilon) = \varepsilon;$$
$$\rho(v\, a) = a\, \rho(v).$$

A longer word, which has been—compared to the former situation—enlarged by 1 character, gets reversed by replacing this character from the end to the front and then still reversing the currently remaining part of the string via the same principle. Generally, the following statement holds:

$$\rho(u\, v) = \rho(v)\, \rho(u).$$

Proof: Induction on the le*n*gth $n$ of a string:

11 We provide preliminary work for the theme of *Formal Languages* in *Theoretical Informatics*.

12 Similar to the order of one's first and last name.

(a) Basis: $|w| := 0\ [= n_0; w := \varepsilon]$

$$\text{Rule: } \rho(\varepsilon) = \rho(\varepsilon\,\varepsilon) = \rho(\varepsilon)\,\rho(\varepsilon),$$

$$\text{Principle: } \rho(\varepsilon\,\varepsilon) = \rho(\varepsilon) = \varepsilon = \varepsilon\,\varepsilon = \rho(\varepsilon)\,\rho(\varepsilon);$$

$$\text{Order: } \rho(u\,v) := \rho(u\,\varepsilon) = \rho(u) = \varepsilon\,\rho(u) = \rho(\varepsilon)\rho(u) = \rho(v)\,\rho(u).$$

(b) Hypothesis:

$$\rho(c_1c_2\ldots c_n) = c_n\ldots c_2c_1;$$
$$\rho(u\,v) = \rho(v)\,\rho(u).$$

(c) Step: $(n_0 \le |v| =)\ n \overset{\text{here}}{\longrightarrow} n+1\ \ (= |va| > n_0)$

$$w := v\,a;$$
$$\begin{aligned}\rho(u\,w) &= \rho(u\,(v\,a)) = \rho((u\,v)\,a) = a\,\rho(u\,v)\\ &\overset{!}{=} a\,(\rho(v)\,\rho(u)) = (a\,\rho(v))\,\rho(u)\\ &= \rho(v\,a)\,\rho(u) = \rho(w)\,\rho(u).\end{aligned}$$

We could prove that each string can get reversed by swapping the reversed part at the rear with the reversed one at the front—independent of the length of the string. ☺

## 4.2 Direct proof

In this approach, one starts with a guaranteed start situation, performs some valid steps (typically equivalence transformations), in order to *directly* yield the statement. To ease the comparison of different proof techniques, we apply the current principle of the *direct proof* on those examples already presented in the preceding section.

**1st illustration:**
This statement is mentioned in Subsection 4.1.1, 1st example (p. 34):

$$e_n = \frac{n\cdot(n-1)}{2}.$$

Proof: Direct:

$$\begin{array}{rl} e_n = e_{n-1} + (n-1) = & 1+2+3+\cdots+(n-3)+(n-2)+(n-1);\\ +\,= & \underline{[(n-1)+(n-2)+(n-3)+\cdots+3+2+1]}\\ 2\cdot e_n = n\cdot(n-1) & |:2 \end{array}$$

$$\iff \quad e_n = \frac{n\cdot(n-1)}{2}.$$

(This formula, to sum up the first $n-1$ natural numbers, is very popular.)

**2nd illustration:**
This statement is already known from Subsection 4.1.1, 2nd example (p. 35):

$$s_{b_{[>1]}}(l) := \sum_{i:=0}^{l} b^i = \frac{b^{(l+1)} - 1}{b - 1}.$$

Proof: Direct:

$$\begin{array}{rcll} 1 \cdot s_b(l) &=& b^0 + b^1 + b^2 + \cdots + b^{(l-1)} + b^l & | \cdot b \\ -[b \cdot s_b(l) &=& b^1 + b^2 + b^3 + \cdots + b^l + b^{(l+1)}] & \\ \hline (1-b) \cdot s_b(l) &=& 1 - b^{(l+1)} \quad | : (1-b) & \end{array}$$

$$\iff \quad s_b(l) = \frac{1 - b^{(l+1)}}{1 - b} \qquad | \cdot \frac{-1}{-1}$$

$$\iff \quad s_b(l) = \frac{b^{(l+1)} - 1}{b - 1}.$$

Here, the key idea is to represent the individual parts appropriately, to further perform an obvious ⌣̈ mathematical operation, followed by a few further steps—and that's it.

## 4.3 Indirect proof

This philosophy is based on the "contrapositive" equivalence in *Boole*an algebra:

$$a \to c \quad \iff \quad \neg c \to \neg a;$$

the truth table in Figure 4.4 illustratively shows the validity of this principle.

| $a$ | $c$ | $a \to c$ | $\neg c \to \neg a$ | $\neg c$ | $\neg a$ |
|---|---|---|---|---|---|
| 0 | 0 | 1 | 1 | 1 | 1 |
| 0 | 1 | 1 | 1 | 0 | 1 |
| 1 | 0 | 0 | 0 | 1 | 0 |
| 1 | 1 | 1 | 1 | 0 | 0 |

**Figure 4.4:** Indirect proof.

We start by negating the propositional variable $c$—in the hope that, after correct transformation steps, to end up with the negation of the original left-hand side ($\neg a$)—a so-called "proof by contradiction". However, like in the *direct* proof, $a$ is considered to be true; hence, the source for its negation could only be the wrong supposition $\neg c$—the reason why $c$ should be true, which has been shown in this *indirect* way.

*Example*: Squaring a natural number is invariant w. r. t. its parity (to be *even*/*odd*), and the square-root output of a square number has the same parity as its input.

Preliminaries to the parity $p$:

- $p(2i) = even \quad \neq \quad odd = p(2i+1)$;
- $p(i) = even \quad \oplus \quad p(i) = odd; \ \forall i \in \mathcal{N}$.

(We consider it as a function; by doing so, each natural number has exactly 1 parity.)

For each arbitrary $n \in \mathcal{N}$, the following statement (already mentioned textually) holds:

$$p(n^2) = p(n).$$

We split the proof:

1. $p(n^2) = \text{even} \implies p(n) = \text{even}$,
2. $p(n^2) = \text{odd} \implies p(n) = \text{odd}$;
3. $p(n) = \text{even} \implies p(n^2) = \text{even}$,
4. $p(n) = \text{odd} \implies p(n^2) = \text{odd}$.

We can abandon the second half; we now treat the first two cases:

Proof: Indirect:

1. To show: $p(n^2) = \text{even} \implies p(n) = \text{even}$:

$$\begin{aligned}
&p(n) \neq \text{even} \quad [p(n) = \text{odd}] \quad \implies \quad n := 2i+1;\\
&p(n^2) = p((2i+1)^2) = p(4i^2+4i+1) = p(2\cdot(2i^2+2i)+1)\\
&\qquad = [p(1) =] \quad \text{odd} \neq \text{even}_{\text{start situation}} \quad \implies\\
&p(n) = \text{odd} \neq \text{even} \quad \implies \quad p(n^2) = \text{odd} \neq \text{even}\\
&\overset{\text{indirect proof}}{\iff}\\
&p(n^2) = \text{even} \quad \implies \quad p(n) = \text{even}.
\end{aligned}$$

2. To show: $p(n^2) = \text{odd} \implies p(n) = \text{odd}$:

$$\begin{aligned}
&p(n) \neq \text{odd} \quad [p(n) = \text{even}] \quad \implies \quad n := 2i;\\
&p(n^2) = p((2i)^2) = p(4i^2) = p(2\cdot 2i^2)\\
&\qquad = [p(0) =] \quad \text{even} \neq \text{odd}_{\text{start situation}} \quad \implies\\
&p(n) = \text{even} \neq \text{odd} \quad \implies \quad p(n^2) = \text{even} \neq \text{odd}\\
&\overset{\text{indirect proof}}{\iff}\\
&p(n^2) = \text{odd} \quad \implies \quad p(n) = \text{odd}\,.
\end{aligned}$$

# 5 Counting techniques

We now reach the highlight of the final part of our book. At the beginning, we enlighten some foundational techniques like the rules of sum, product and quotient[1] as well as the pigeonhole principle. Then we consider the inclusion/exclusion of set expressions, mainly to compute the number of elements in the union of nondisjunctive sets (with nonempty intersections). We proceed with the recurrence relation (my favorite ☺). Thereby, I like to take you with me on a creativity trip: the main purpose of this approach is to creatively find a closed formula. Afterwards, we treat the classics among the counting problems: the number of different orders (permutations) and selections (combinations); thereby, we distinguish between objects of different and identical types. Finally, we discuss the Stirling numbers of the first and second kind, basically in their interpretation of cycle and subset numbers, and the Bell numbers. (Further items are treated in my Informatics book,[2] for instance, in the 1st chapter, where in Section "1.4 Zähltechniken" [pp. 14–16], among others, we illustrate "selection with repetition" by visiting a night bar ☺. [There, an introduction to *cryptology* follows, including the *Eulerean* $\phi$-function, Euler–Fermat's little ☺ theorem, as well as its use in the special case of a *Prime* number in the *RSA*-procedure.])

## 5.1 Basics

### 5.1.1 Rule of sum

Given are $m$ different cases à $n_i$ $(1 \le i \le m)$ different options; then the number of different possibilities is

$$z := \sum_{i:=1}^{m} n_i.$$

Let us consider an (unusually small) example from Informatics:

The *l*ength $l$ of a user name should range between 1 and 3; for $l := 1$ we should use a digit ($\in \{0, 1, 2, \ldots, 9\}$) or a vocal ($\in \{a, e, i, o, u\}$), for $l := 2$, we must place a digit in front of it, and for $l := 3$ we are forced to place a vocal at the now very left position.

*Question:* How many possibilities for the formation of such a user name[3] do exist?

1 Not to mismatch the notions with those in differential analysis.

2 Still just in German.

3 Admittedly rather unsecure.

https://doi.org/10.1515/9783111206899-006

*Answer:*

$$z :=_{\leftarrow} \sum_{i:=1}^{3} n_i = \big|\{0,1,2,\dots,9\} \cup \{\mathtt{a},\mathtt{e},\mathtt{i},\mathtt{o},\mathtt{u}\}\big| + 10 \cdot 15 + 5 \cdot 150$$
$$= 15 \cdot (1 + 10 + 5 \cdot 10) = 15 \cdot 61 = 915.$$

*Additional question:* Which $z$ do we yield once we develop the string from the other side, i. e., for $l := 1$, we start with a vocal, for $l := 2$, we place a digit behind and for $l := 3$ we end up with a digit or vocal? Before starting to compute it: Is it not anyhow the same?

*Additional answer:*

$$z :=_{\rightarrow} \sum_{i:=1}^{3} n_i = 5 + 5 \cdot 10 + 50 \cdot 15$$
$$= 5 \cdot (1 + 10 + 10 \cdot 15) = 5 \cdot 161 = 805 \neq_{\smile} 915.$$

One of the 3 sum*m*ands is, of course, identical with the corresponding one in the first construction above. Which one is it?[4]

You find a more complex enlargement in my Informatics book (pp. 16–21) cited at the end.

### 5.1.2 Rule of product

Given are $m$ (different) steps (positions) à $n_i$ ($1 \le i \le m$) different options (assignments, resp.); then the number of different possibilities is

$$z := \prod_{i:=1}^{m} n_i.$$

Let us consider a standard example from Informatics:
the number of coding possibilities of an $m$-ary bit-vector.[5]

*Question:* How many possibilities of such a string do exist?

*Answer:*

$$z := \prod_{i:=1}^{m} n_i = |\mathcal{B}|^m = 2^m \quad [= |\mathcal{B}^m|].$$

---

4 $n_m$ ($=_{\text{here}}$ $n_3 = 50 \cdot 15 = 5 \cdot 150 = 750$).

5 Binary string with predefined length $l := m$ and assignment options at each position from the 2-elementary set $\mathcal{B} := \{0, 1\}$.

### 5.1.3 Rule of quotient

Given is a division of an $n$-elementary set into subsets of equal size à $(0 <)$ $m$ $(< n)$ elements; then the number of such subsets is

$$z := \frac{n}{m}.$$

Let us consider an interesting example from the world of permutations (see p. 69): the number of orders to place $m$ different elements.

*Question:* How many different possibilities of placements[6] of $m$ persons do exist on a round table, whereby the position labeling of the table is unimportant?

(Here, we already use the !-sign for the "factorial"; you might regard the final recurrence example presented later as well as the coming section "Permutation".)

*Question:* $z := n/m = m!/m = (m-1)! \cdot m/m = (m-1)!$.

*Explanation:* $S :=$ set of all principally possible orderings of $m$ persons; $|S| = m! =: n$.

Out of these $n$ different permutations $m$ of them have a common round-table ordering (according to the placement rule of having the same neighboring situation)—the reason why exactly 1 representative suffices to describe this pattern. Then

$$z := n/m$$

indicates the number of different representatives—each one staying for a subset of all those permutations, which are just cyclically shifted patterns.

### 5.1.4 Pigeonhole principle

This very simple, nevertheless helpful, idea[7] is as follows: $p$ pigeons fly to $h$ holes; then there exists at least 1 hole with at least this number $z$ of pigeons:

$$z := \left\lceil \frac{p}{h} \right\rceil.$$

*Example:* Exam organization

$$p := \#\text{ students},$$
$$h := \#\text{ (exam) rooms}.$$

6 Related to "is located exactly 1 position left (or right) of the neighbor".

7 From 1834 by Johann Peter Gustav Lejeune Dirichlet.

Then it is not possible that in each room there are less than $z$ students; formulated positively: there exists at least 1 room with at least $z := \lceil p/h \rceil$ students.

*Illustration:* Given the two values $p := 65$ and $h := 3$.

*Question:* Which number do we get for $z$?

*Answer:*

$$z := \left\lceil \frac{65}{3} \right\rceil = \left\lceil \frac{3 \cdot 21 + 2}{3} \right\rceil = \left\lceil 21 + \frac{2}{3} \right\rceil = 22.$$

*Interpretation:* It is not sufficient to place only 21 chairs in each room; at least in one room we must have at least 22.

## 5.2 In/exclusion

Here, we treat the counting of elements in the union of potentially nondisjunctive[8] sets. At first, we indicate for $n$ sets the number of nonempty combinations[9]:

$$\sum_{i:=1}^{n} \binom{n}{i} = \sum_{i:=0}^{n} \binom{n}{i} - \binom{n}{0} = 2^n - 1 =: c.$$

Then, the # elements $u$ in the set union is as follows, thereby considering $c$ terms:

$$u = \left| \bigcup_{k:=1}^{n} A_k \right| = \sum_{k:=1}^{n} \left( (-1)^{(k+1)} \cdot \sum_{1 \le i_1 < \cdots < i_k \le n} | \cap A_{i_j} | \right).$$

At first glance, this looks a little bit cumbersome which, however, becomes (easily) clear by the following concretizations:

(i) $n := 1$:

*Construction:*

$$\bigcup_{k:=1}^{1} A_k = A;$$

*Counting formula:*

$$\left| \bigcup_{k:=1}^{1} A_k \right| = \sum_{k:=1}^{1} \left( (-1)^{(k+1)} \cdot \sum_{1 \le i_1 < \cdots < i_k \le 1} | \cap A_{i_j} | \right)$$
$$= (-1)^{(1+1)} \cdot | \cap A_1 | = (-1)^2 \cdot | \cap A | = |A|.$$

8 Perhaps having at least one common element.

9 The following notation $\binom{n}{k}$, sometimes written as $C(n, k)$, gets introduced in the Subsection 5.4.2 (starting from p. 72): # possibilities to pick $k$ elements from an $n$-elementary set.

(ii) $n := 2$:
*Construction:*

$$\bigcup_{k:=1}^{2} A_k = A_1 \cup A_2;$$

*Counting formula:*

$$\begin{aligned}\left|\bigcup_{k:=1}^{2} A_k\right| &= \sum_{k:=1}^{2}\left((-1)^{(k+1)} \cdot \sum_{1\le i_1<\cdots<i_k\le 2} |\cap A_{i_j}|\right)\\ &= (-1)^{(1+1)} \cdot (|A_1| + |A_2|) + (-1)^{(2+1)} \cdot |A_1 \cap A_2|\\ &= |A_1| + |A_2| - |A_1 \cap A_2|.\end{aligned}$$

(iii) $n := 3$:
*Construction:*

$$\bigcup_{k:=1}^{3} A_k = A_1 \cup A_2 \cup A_3;$$

*Counting formula:*

$$\begin{aligned}\left|\bigcup_{k:=1}^{3} A_k\right| &= \sum_{k:=1}^{3}\left((-1)^{(k+1)} \cdot \sum_{1\le i_1<\cdots<i_k\le 3} |\cap A_{i_j}|\right)\\ &= (-1)^{(1+1)} \cdot (|A_1| + |A_2| + |A_3|)\\ &\quad + (-1)^{(2+1)} \cdot (|A_1 \cap A_2| + |A_1 \cap A_3| + |A_2 \cap A_3|)\\ &\quad + (-1)^{(3+1)} \cdot |A_1 \cap A_2 \cap A_3|\\ &= |A_1| + |A_2| + |A_3| - |A_1 \cap A_2| - |A_1 \cap A_3| - |A_2 \cap A_3| + |A_1 \cap A_2 \cap A_3|.\end{aligned}$$

Two examples might illustrate the application of this counting principle:

a) A football manager has to handle this situation:

$S :=$ set of all players in the squad ($S := D \cup F \cup N$, see below), $|S| := 13$;

$D :=$ set of players suitable for the *d*efense, $|D| := 9$;

$F :=$ set of players suitable for the *o*ffense, $|F| := 6$;

$N :=$ set of *n*onplayers (too bad or injured), $|N| := 2$;

$M :=$ set of *m*idfielders $:=_{\mathtt{here}} D \cap F$,

$D_o :=$ defenders$_{\mathtt{only}}$,

$F_o :=$ strikers$_{\mathtt{only}}$.

W. r. t. the line-up, the manager is interested in the answering of this.

*Questions:*
Do we have enough midfielders?
How many players could just play on the defense, how many pure strikers do we have?

*Answer:*

$$
\begin{aligned}
|D \cup F| &= |D| + |F| - |D \cap F| \quad \Longleftrightarrow \\
|D \cap F| &= |D| + |F| - |D \cup F| = 9 + 6 - (|S| - |(D \cup F)^c|) = 15 - (13 - |N|) \\
&= 2 + 2 = 4 = |M|; \\
|D_o| &= |D \setminus M| =_{[D \supseteq M]} |D| - |M| = 9 - 4 = 5, \\
|F_o| &= |F \setminus M| =_{[F \supseteq M]} |F| - |M| = 6 - 4 = 2.
\end{aligned}
$$

The goalie is part of the defense; here, the game can begin: $5 + 4 + 2 = 11$.
We could yield the number of midfielders slightly more elegant:

$$
\begin{aligned}
|S| := |(D \cup F) \cup N| &\overset{\text{partition}}{\underset{\text{on } 2^{\text{nd}} \cup}{=}} (|D| + |F| - |D \cap F|) + |N| \quad \Longleftrightarrow \\
|D \cap F| &= |D| + |F| + |N| - |S| = 9 + 6 + 2 - 13 = 4 = |M|.
\end{aligned}
$$

b) We have the following scenario to plan the number of exam papers to get printed:

$$
\begin{aligned}
DM &:= \text{set of candidates in } D\text{iscrete } M\text{athematics}; |DM| := 60, \\
FL &:= \text{set of candidates in } F\text{ormal } L\text{anguages}; |FL| := 50, \\
AI &:= \text{set of candidates in } A\text{rtificial } I\text{ntelligence}; |AI| := 40; \\
DM \cap FL &:= \text{set of candidates involved in } DM \text{ and in } FL; |DM \cap FL| := 40, \\
DM \cap AI &:= \text{set of candidates involved in } DM \text{ and in } AI; |DM \cap AI| := 30, \\
FL \cap AI &:= \text{set of candidates involved in } FL \text{ and in } AI; |FL \cap AI| := 20, \\
DM \cap FL \cap AI =: C &:= \text{set of candidates, who have to pass all 3 exams}; |C| =: c := 10.
\end{aligned}
$$

$A := DM \cup FL \cup AI$ set of *a*ll candidates overall; $|A| =: a$. Interesting to know is the

*Question:*
How many students are involved in which exam(s):
at least 1, exactly 2 of the 3 or just 1—which ones?

*Answer:*

$$
\begin{aligned}
a &= |DM \cup FL \cup AI| \\
&= |DM| + |FL| + |AI| - (|DM \cap FL| + |DM \cap AI| + |FL \cap AI|) + c \\
&= 60 + 50 + 40 - (40 + 30 + 20) + 10 = 160 - 90 = 70.
\end{aligned}
$$

70 students are involved in at least 1 exam.

$$\begin{aligned}
|(DM \cap FL) \setminus C| &=_{[DM \cap FL \supseteq C]} |DM \cap FL| - c = 40 - 10 = 30.\\
|(DM \cap AI) \setminus C| &=_{[DM \cap AI \supseteq C]} |DM \cap AI| - c = 30 - 10 = 20.\\
|(FL \cap AI) \setminus C| &=_{[FL \cap AI \supseteq C]} |FL \cap AI| - c = 20 - 10 = 10.
\end{aligned}$$

30 (out of 70) students exactly write the two exams $DM$ and $FL$, still 20 students exactly $DM$ and $AI$, and just 10 students exactly $FL$ and $AI$.

$$\begin{aligned}
|DM_{\text{excl.}}| &= |S \setminus (FL \cup AI)| =_{[S \supseteq FL \cup AI]} s - (|FL| + |AI| - |FL \cap AI|)\\
&= 70 - (50 + 40 - 20) = 0.\\
|FL_{\text{excl.}}| &= |S \setminus (DM \cup AI)| =_{[S \supseteq DM \cup AI]} s - (|DM| + |AI| - |DM \cap AI|)\\
&= 70 - (60 + 40 - 30) = 0.\\
|AI_{\text{excl.}}| &= |S \setminus (DM \cup FL)| =_{[S \supseteq DM \cup FL]} s - (|DM| + |FL| - |DM \cap FL|)\\
&= 70 - (60 + 50 - 40) = 0.
\end{aligned}$$

Nobody has just 1 exam to pass; $0 \cdot 3 + (30 + 20 + 10) + c = s$.

## 5.3 Recurrence relation

This technique is mainly used in the case of a counting problem once we do not have a closed formula at hand; i. e., we could not immediately produce for any given input its output value (the result of the corresponding counting challenge). What we, however, can often recognize is the change, which happens when we enlarge the problem by 1 step. In Informatics, for example, having a certain number $n-1$ of bits, let us now enlarge this # bits by exactly 1 to $n$, we would double the number of possible combinations, resulting in $2^n$ combinations overall—a closed formula.[10]

Let us illustrate the idea on the *sum of Gauß*,[11] the simplest *arithmetical*[12] *series*:

$$g(n) := \sum_{i:=1}^{n} i;$$

here, the closed formula $g_n = n \cdot (n+1)/2$ should still be unknown for the moment.[13] We easily observe that from step $n-1$ to $n_{[>0]}$ exactly $n$ gets added to $g_{n-1}$; hence, the

10 For example, it would not be necessary to double the 1 ($= 2^0$) 8 times to yield the 256 ($= 2^8$), because we are now aware of this exponential dependency.

11 Initially, more than 200 years ago, formulated by his teacher as an incremental procedure.

12 $a_0 := 0$, $a_i :=_{[i>0]} a_{i-1} + d_{\underline{d}\text{ifference}}$; $\sum_{i:=0}^{n} a_i =_{\text{here}} \sum_{i:=1}^{n} a_i = \sum_{i:=1}^{n} d \cdot i = d \cdot [n \cdot (n+1)/2] = n \cdot (n+1)/2$.

13 Sometimes, instead of $o(i)$ we use $o_i$ to ease reading, especially in cases where the input represents a composed expression, like $i-1$, to avoid the interpretation of performing a multiplication during the production of the output. Traditionally, $o(i-1)$ could be used; however, it should not be read as $o \cdot (i-1)$; therefore, it is recommendable to notate the output $o_{i-1}$ w. r. t. the input $i-1$.

following recursion holds:

$$g_n := g_{(n-1)} + n;$$

i. e., we know the incremental construction of $g_n$.

We can now apply this principle to the predecessor case:

$$g_{(n-1)} := g_{(n-2)} + (n-1),$$

and thereby

$$g_n := g_{(n-2)} + (n-1) + n,$$

then

$$g_n := g_{(n-3)} + (n-2) + (n-1) + (n-0) := \cdots .$$

How many backward steps are needed? We go $n-1$ steps, "down to" $g_1$ (:= 1), and get

$$g_n = 1 + 2 + 3 + \cdots + (n-2) + (n-1) + n.$$

This approach is called "backward substitution".

Starting by

$$g_1 := 1,$$

producing

$$g_2 := g_1 + 2 = 1 + 2 = 3,$$
$$\vdots$$
$$g_n := g_{(n-1)} + n = \cdots,$$

we end up (of course)—after $n-1$ steps—with the same sum for $g_n$; this procedure is called "forward substitution". We collect the objects as follows: the first element, the 1, with the last element $n$, the second element, 2, with the second from last element $n-1$, the third element, 3, with the third from last element $n-2$, etc. This works up to the middle $\lfloor \frac{n}{2} \rfloor$; once $n$ is odd, we must add $\lceil \frac{n}{2} \rceil$ exactly 1 times.[14] We obtain:

**14** Surely not twice—what could happen for an odd $n$, if in the following part "We obtain" we had wrongly used the ceiling rounding function for the upper index in the sum without case analysis related to the parity ("even" or "odd").

$$\begin{aligned}
g_n &= \sum_{i:=1}^{\lfloor \frac{n}{2} \rfloor} (i + (n - (i-1))) + \begin{cases} 0; & \text{even}(n) \\ \lceil \frac{n}{2} \rceil; & \text{odd}(n) \end{cases} \\
&= \begin{cases} \sum_{i:=1}^{\frac{n}{2}} (i + n - i + 1); & \text{even}(n) \\ \sum_{i:=1}^{\frac{n-1}{2}} (n+1) + \lceil \frac{n}{2} \rceil; & \text{odd}(n) \end{cases} \\
&= \begin{cases} \frac{n}{2} \cdot (n+1); & \text{even}(n) \\ \frac{n-1}{2} \cdot (n+1) + \frac{n+1}{2} = ((n-1)+1) \cdot \frac{n+1}{2}; & \text{odd}(n) \end{cases} \\
&= \frac{n \cdot (n+1)}{2} \quad ☺.
\end{aligned}$$

A usual induction proof would finally confirm the statement.[15]

This sounds a little bit strange, especially once we are aware of the Gaussean sum formula. We are still in the warm-up phase in order to get an insight into the principal approach how to perform a recurrence relation.

In practical counting, it is often the case that we do not immediately see the direct dependence between input and output, whereas the incremental structure is visible; in this case, it is helpful having the recurrence relation at hand.

Let us collect the fundamental parts:

- Scheme: similar to an “incremental” recursion
- Idea: elimination of this recursion in order to yield a closed formula
- Structure
  - Basic index with value
  - Construction principle
    * using the predecessor value
    * constructive step (creative part)
  - Development/substitution (two versions)
    * backwards
    * forwards
- Proof (of the statement detected—often via similarly built induction).

#### 1st illustration: *n*-ary bit-vector (binary string with length *n*)

Having $n$ array positions, we count in $z_n$ the number of possibilities that in exactly 1 of these cells a *false*[16] bit (here mentioned as “0” [Zero]) is stored. (That $z_n$ obviously results in $n$ is currently irrelevant; we take this task just to illustrate the principal structure of a recurrence.)

15 Actually, the recursion (step) would function as the induction step anyhow; hence, whether a further proof is still needed, I am not sure. ☺

16 *True* (represented by “1”) would produce the same result.

Basic value: $z_0 := 0$;
Principle: $z_{n_{[>0]}} := z_{(n-1)} + 1$.

Why? An $(n-1)$-ary bit-vector could only get enlarged at the front by a "1" (:= *true*) or a "0" (:= *false*) to produce an $n$-ary bit string. What happens once we place a "1" at the front? This case obviously cannot affect the number in question, because we focus on the appearance of a "0"; therefore, all former cases still remain ($z_{n-1}$). What happens once we place a "0" at the front? Only in the (single) case that in all other $n-1$ former positions there would not exist any *false* (i. e., all former entries are *true*), we now form with a new leading "0" a bit-vector with the required structure. Therefore, we obtain, for $n > 0$:

$$z_n := z_{(n-1)} + \binom{n-1}{0},$$

where the addend last mentioned, the "Binomial coefficient",[17] provides the number of possibilities how often we could have exactly 0 times *false* in $n-1$ bit positions—out of $2^{(n-1)}$ *Boole*'s assignment possibilities, this is exactly 1 times the case (please refer to the aforementioned "principle"). Figure 5.1 would like to illustrate exactly that.

**Figure 5.1:** Bit-pattern recurrence.

17 Gets introduced in the Subsection 5.4.2 (starting on p. 72).

We now demonstrate that both substitution versions yield the intended formula:

- Backward substitution:

$$\begin{aligned} z_n &:= z_{(n-1)} + 1 := (z_{(n-2)} + 1) + 1 \\ &= z_{(n-2)} + 2 := (z_{(n-3)} + 1) + 2 \\ &= z_{(n-3)} + 3 := (z_{(n-4)} + 1) + 3 \\ &= z_{(n-4)} + 4 \\ &\vdots \\ &:= z_{(n-n)} + n \\ &= z_0 + n \\ &:= 0 + n \\ &= n; \end{aligned}$$

- Forward substitution:

$$\begin{aligned} z_1 &:= z_0 + 1 := 0 + 1 = 1 \\ z_2 &:= z_1 + 1 := 1 + 1 = 2 \\ z_3 &:= z_2 + 1 := 2 + 1 = 3 \\ z_4 &:= z_3 + 1 := 3 + 1 = 4 \\ &\vdots \\ z_n &:= z_{(n-1)} + 1 := (n-1) + 1 = n. \end{aligned}$$

Statement: $z_n = n\ [= \binom{n}{1}]$.

Proof: Induction on $n$:

- Basis:[18] $n_0 := 0$
  - Principle: $z_0 = 0$ [Initial value: $0 \times$ *false*]
  - Formula: $z_0 = 0 \mathrel{\hat{=}}$ Principle.
- Hypothesis: $z_{(n-1)} = n - 1$.
- Step: $(n_0 \leq)\, n-1 \rightarrow n\ (> n_0)$
  - Principle: $z_n := z_{(n-1)} + 1 \stackrel{!}{=} (n-1) + 1 = n$
  - Formula: $z_n = n \mathrel{\hat{=}}$ Principle.

18 Alternatively: $n_0 := 1$
- Principle: $z_1 = 1$
- Formula: $z_1 = 1$

**2nd illustration: # edges in a "complete" graph with *n* nodes**

We count by $e_n$ the number of undirected edges in a graph, where each of the $n_{[>0]}$ nodes is connected exactly once with each of the other $n-1$ nodes.

Basic value: $e_1 := 0$;
Principle: $e_n := e_{(n-1)} + (n-1)$.

Why? A complete graph with $n-1$ nodes can retain this feature remaining a complete graph during an extension by 1 node only when the new $n^{\text{th}}$ node, in addition to the $e_{(n-1)}$ edges already there,[19] gets connected by 1 individual edge to each of the $n$–1 nodes already existing—the reason for the construction principle presented above.

We show that both substitution versions produce the closed formula:

- Backward substitution:

$$
\begin{aligned}
e_n &:= e_{(n-1)} + (n-1) \\
&:= e_{(n-2)} + (n-2) + (n-1) \\
&:= e_{(n-3)} + (n-3) + (n-2) + (n-1) \\
&\vdots \\
&:= e_{(n-(n-1))} + \big(n-(n-1)\big) + \cdots + (n-3) + (n-2) + (n-1) \\
&= e_1 + \sum_{i:=1}^{n-1} i \\
&:= 0 + \sum_{i:=1}^{n} i - n \\
&= \frac{n\cdot(n+1)}{2} - \frac{n\cdot 2}{2} \\
&= \frac{n\cdot((n+1)-2)}{2} \\
&= \frac{n\cdot(n-1)}{2};
\end{aligned}
$$

- Forward substitution:

$$
\begin{aligned}
e_2 &:= e_1 + 1 := 0 + 1 = 1 \\
e_3 &:= e_2 + 2 := 1 + 2 = 3 \\
e_4 &:= e_3 + 3 := 3 + 3 = 6 \\
&\vdots \\
e_n &= \sum_{i:=0}^{n-1} i =_{\substack{\texttt{according} \\ \texttt{to Gauß}}} \frac{(n-1)\cdot n}{2}.
\end{aligned}
$$

19 No edge should get deleted.

Statement: $e_n = \frac{n\cdot(n-1)}{2}$.

Proof: See Subsection 4.1.1, 1st example (p. 34).

**3rd illustration: # edges in an *h*-dimensional hypercube with $n_h$ nodes**

A so-called "*h*ypercube" in the dimension $h_{[>0]}$ with exactly 1 *n*ode at each of the $n_h$ ($= 2^h$) corners is constructible from its predecessor structure with dimension $h-1$ by doubling this $h-1$ thing and additionally placing 1 further edge from each corner to the corresponding corner of the clone as depicted in Figure 5.2. In doing so, the number of corners gets doubled ($n_h = 2 \cdot n_{h-1}$); expressed differently: the hypercube of dimension $h-1$ has half of the # nodes as the *h*-dimensional one: $n_{h-1} = n_h/2$. However, as in the preceding example, we are interested in the # *c*onnections ($=: c_h$). To initiate a recursion, we need a parameter which is incrementable by 1: hence, the dimension plays the role of the recurrence relation.

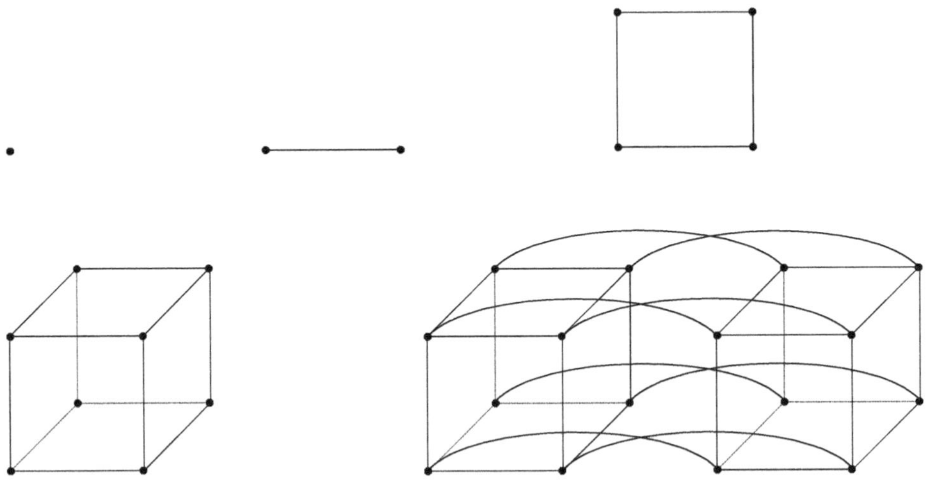

**Figure 5.2:** *h*-dimensional hypercube.

Basic value: $c_0 := 0$;
Principle: $c_h := 2 \cdot c_{h-1} + n_{h-1} = 2 \cdot c_{h-1} + 2^{h-1}$.

Why? By the enlargement of the dimension (from $h-1$ to $h$), we first build a copy of the hypercube with dimension $h-1$ and thereby already obtain in this initial stage double the number related to the # former nodes; furthermore, we add for each node in the former structure of 1 dimension less 1 new edge. Therefore, we obtain, for $h > 0$, the aforementioned recurrence principle.

We now show that both substitution versions produce the closed formula:

– Backward substitution:

$$\begin{aligned}
c_h &:= 2 \cdot c_{h-1} + 2^{h-1} \\
&:= 2 \cdot (2 \cdot c_{h-2} + 2^{h-2}) + 2^{h-1} \\
&= 4 \cdot c_{h-2} + 2^1 \cdot 2^{h-2} + 2^{h-1} \\
&= 4 \cdot c_{h-2} + 2^{h-1} + 2^{h-1} \\
&:= 4 \cdot (2 \cdot c_{h-3} + 2^{h-3}) + 2 \cdot 2^{h-1} \\
&= 8 \cdot c_{h-3} + 2^2 \cdot 2^{h-3} + 2 \cdot 2^{h-1} \\
&= 8 \cdot c_{h-3} + 1 \cdot 2^{h-1} + 2 \cdot 2^{h-1} \\
&= 2^3 \cdot c_{h-3} + 3 \cdot 2^{h-1} \\
&:= 2^3 \cdot (2 \cdot c_{h-4} + 2^{h-4}) + 3 \cdot 2^{h-1} \\
&= 2^4 \cdot c_{h-4} + 1 \cdot 2^{h-1} + 3 \cdot 2^{h-1} \\
&= 2^4 \cdot c_{h-4} + 4 \cdot 2^{h-1} \\
&\vdots \\
&:= 2^h \cdot c_{h-h} + h \cdot 2^{h-1} \\
&= n_h \cdot c_0 + h \cdot 2^{h-1} \\
&:= h \cdot 2^{h-1};
\end{aligned}$$

– Forward substitution:

$$\begin{aligned}
c_1 &:= 2 \cdot c_0 + 2^0 := 2 \cdot 0 + 2^0 = 1 \cdot 2^0 \\
c_2 &:= 2 \cdot c_1 + 2^1 := 2 \cdot 2^0 + 2^1 = 2 \cdot 2^1 \\
c_3 &:= 2 \cdot c_2 + 2^2 := 2 \cdot 2^2 + 1 \cdot 2^2 = 3 \cdot 2^2 \\
c_4 &:= 2 \cdot c_3 + 2^3 := 2 \cdot (3 \cdot 2^2) + 2^3 = 3 \cdot 2^3 + 1 \cdot 2^3 = 4 \cdot 2^3 \\
c_5 &:= 2 \cdot c_4 + 2^4 := 2 \cdot (4 \cdot 2^3) + 2^4 = 4 \cdot 2^4 + 1 \cdot 2^4 = 5 \cdot 2^4 \\
&\vdots \\
c_h &:= h \cdot 2^{h-1}
\end{aligned}$$

Statement: $c_h = h \cdot 2^{h-1}$.

Proof: Induction on $h$ [$= \mathtt{log}_2(2^h) =:_{\underline{\mathtt{dualis}}}^{\mathtt{logarithmus}} \mathtt{ld}(n_h)$]:
- Basis: $h_0 := 0$
  - Principle: $c_0 = 0$ [the 0-dimension point does not have any connection]
  - Formula: $c_0 = 0 \cdot \ldots = 0 \mathrel{\hat{=}}$ principle.
- Hypothesis: $c_{(h-1)} = (h-1) \cdot 2^{(h-1)-1}$.

- Step: $(h_0 \leq) h-1 \rightarrow h (> h_0)$
  - Principle: $c_h := 2 \cdot c_{h-1} + 2^{h-1}$

$$\overset{!}{=} 2^1 \cdot ((h-1) \cdot 2^{(h-1)-1}) + 1 \cdot 2^{h-1}$$
$$= ((h-1)+1) \cdot 2^{h-1}$$
$$= h \cdot 2^{h-1}$$

  - Formula: $c_h = h \cdot 2^{h-1} \triangleq$ Principle.

Result: $c_h(n_h) = \mathrm{ld}(n_h) \cdot 2^h \cdot 2^{-1} = h \cdot n_h/2$.

Of course, ☺: In the $h$-dimensional hypercube, each of the $n_h$ corners has $h$ edges to each its $h$ neighbors, principally; however, due to the fact that none of these edges are directed (an edge does not exist twice), we have to halve this product mentioned before.

**4th illustration: Diameter in the quadratic grid with *n* nodes**

The "diameter" of a graph is the distance (:= length of the shortest path possible) of those two nodes, which are farthest away from each other—along the given topology (edges presented). Here, we work on a quadratic grid structure. Figure 5.3 shows a (5×5)-pattern: # nodes =: $n_5 := 5^2 = 25$. Generally: $n_r = r^2$, $r := \sqrt{n_r}$ := # rows as well as # columns. This $r$ now naturally plays the role of the recurrence parameter; see the following explanation:

Basic value: $d_1 := 0$;
Principle: $d_r := d_{r-1} + 1 \cdot 2$.

Why? Due to the enlargement by 1 further row and 1 further column from $r$–1 to $r$ right now we must add to the former distance just 1 further step in each of the 2 dimensions. Therefore, we obtain, for $r > 1$, the aforementioned recurrence principle.

We show that both substitution versions yield the closed formula:

- Backward substitution:

$$d_r := d_{r-1} + 1 \cdot 2$$
$$:= (d_{r-2} + 1 \cdot 2) + 1 \cdot 2 = d_{r-2} + 2 \cdot 2$$

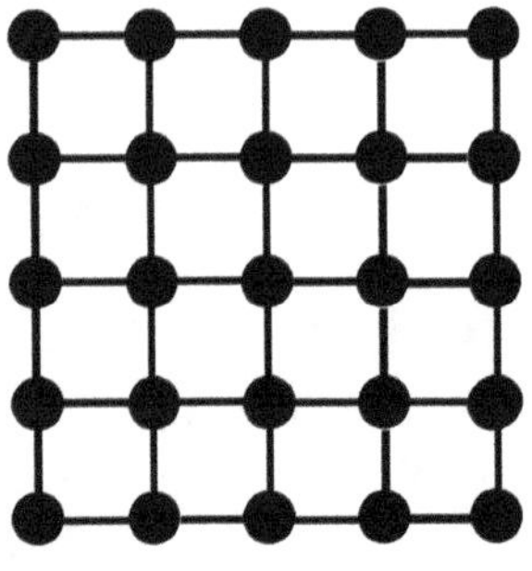

**Figure 5.3:** Grid.

$$
\begin{aligned}
&:= (d_{r-3} + 1 \cdot 2) + 2 \cdot 2 = d_{r-3} + 3 \cdot 2 \\
&:= (d_{r-4} + 1 \cdot 2) + (4 - 1) \cdot 2 = d_{r-4} + 4 \cdot 2 \\
&:= (d_{r-5} + 1 \cdot 2) + (5 - 1) \cdot 2 = d_{r-5} + 5 \cdot 2 \\
&:= (d_{r-6} + 1 \cdot 2) + (6 - 1) \cdot 2 = d_{r-6} + 6 \cdot 2 \\
&\vdots \\
&:= (d_{r-(r-1)} + 1 \cdot 2) + ((r-1) - 1) \cdot 2 = d_1 + (r-1) \cdot 2 \\
&:= 2 \cdot (r-1);
\end{aligned}
$$

- Forward substitution:

$$
\begin{aligned}
d_2 &:= d_1 + 1 \cdot 2 := 0 + 1 \cdot 2 = 1 \cdot 2 = (2-1) \cdot 2 \\
d_3 &:= d_2 + 1 \cdot 2 := 1 \cdot 2 + 1 \cdot 2 = 2 \cdot 2 = (3-1) \cdot 2 \\
d_4 &:= d_3 + 1 \cdot 2 := 2 \cdot 2 + 1 \cdot 2 = 3 \cdot 2 = (4-1) \cdot 2 \\
d_5 &:= d_4 + 1 \cdot 2 := 3 \cdot 2 + 1 \cdot 2 = 4 \cdot 2 = (5-1) \cdot 2 \\
d_6 &:= d_5 + 1 \cdot 2 := 4 \cdot 2 + 1 \cdot 2 = 5 \cdot 2 = (6-1) \cdot 2 \\
&\vdots \\
d_r &:= d_{r-1} + 1 \cdot 2 = ((r-1) - 1) \cdot 2 + 1 \cdot 2 = (r-1) \cdot 2
\end{aligned}
$$

Statement: $d_r = 2 \cdot (r-1)$.

Proof: Induction on $r$ [$= \sqrt{n_r}$]:
- Basis: $r_0 := 1$
  - Principle: $d_1 = 0$ [no neighborhood cells: 0 steps]
  - Formula: $d_1 = 2 \cdot (1-1) = 0 \mathrel{\hat{=}}$ Principle.
- Hypothesis: $d_{(r-1)} = 2 \cdot ((r-1) - 1)$.
- Step: $(r_0 \leq) r - 1 \to r (> r_0)$
  - Principle: $d_r := d_{r-1} + 1 \cdot 2$

$$
\begin{aligned}
&\overset{!}{=} 2 \cdot ((r-1) - 1) + 2 \cdot 1 \\
&= 2 \cdot (((r-1) - 1) + 1) \\
&= 2 \cdot (r-1)
\end{aligned}
$$

  - Formula: $d_r = 2 \cdot (r-1) \mathrel{\hat{=}}$ Principle.

Result: $d_r = 2 \cdot (r-1) \;=\; 2 \cdot (\sqrt{n_r} - 1) =: d_{n_r}$.

"Unfortunately", we are in a corner and like to go to that corner farthest away from it; in each of the 2 dimensions we have to make $r-1$ steps, altogether $2 \cdot (\sqrt{n_r} - 1)$.

Applications (on the *diameter* in the current illustration):
- During the routing of an email in a grid it is imaginable that the recurrence principle could be of some interest: On the way to enlarge the network by 1 column and 1 row,

the net admin might reflect the # steps, which are then needed, in addition to the # steps in the smaller grid ("worst case" consideration)—here just 2. This incremental number of additional steps is, of course, independent of the size of the network in mind.

- Once we like to realize, in game design, the shortest path from one corner to that one farthest away from it (to move a joy stick traversing in parallel to each axis), on such a quadratic grid with $r \times r$ access points we need $d_r$ steps to the goal point on an optimal path. Therefore, we are interested in the closed formula $d_r$, obtainable by a/the recurrence relation.

### 5th illustration: # connections in the quadratic ($r \times r$) grid

As in the 3rd illustration, we count the connections in the network architecture; now we use again the input size $r$ as above, the indicator for the # columns (respectively rows) in the given matrix. Finally, we enrich this reference to end up by the actual input parameter $n_r$, the # nodes (which are functionally related to this $r$) in the network.

Basic value: $c_0 := 0;$
Principle: $c_r := c_{r-1} + 4 \cdot (r-1).$

Why? Let us consider the figure with the grid. Once we like to introduce a new $r^{\text{th}}$ row, we must add to the existing $r-1$ nodes in the preceding row (which has, due to its quadratic structure, exactly so many columns and, therefore, these $r-1$ entries in this row) to each of these nodes 1 further vertical connection and along this new row to connect these $r-1$ new nodes horizontally, which gets done by $r-2$ further connections. Finally, we connect the new corner node in the horizontal dimension. Due to the fact that everything in 1 dimension happens (symmetrically) in the other dimension as well, we just have to double these operations at the very end—by which we obtain, for $r > 0$, the recurrence principle:

$$a := \text{additional connections:}$$
$$a_r = 2 \cdot [(r-1) + (r-2) + 1] = 2 \cdot (2 \cdot r - 2) = 4 \cdot (r-1) = a_{n_r} = 4 \cdot \sqrt{n_{r-1}}.$$

We now show that both substitution versions yield the closed formula:

- Backward substitution:

$$\begin{aligned}
c_r &:= c_{r-1} + 4 \cdot (r-1) \\
&:= c_{r-2} + 4 \cdot (r-2) + 4 \cdot (r-1) = c_{r-2} + 8 \cdot r - 12 \\
&:= c_{r-3} + 4 \cdot (r-3) + 8 \cdot r - 12 = c_{r-3} + 12 \cdot r - 24 \\
&:= c_{r-4} + 4 \cdot (r-4) + 12 \cdot r - 24 = c_{r-4} + 16 \cdot r - 40 \\
&:= c_{r-5} + 4 \cdot (r-5) + 16 \cdot r - 40 = c_{r-5} + 20 \cdot r - 60 \\
&:= c_{r-6} + 4 \cdot (r-6) + 20 \cdot r - 60 = c_{r-6} + 24 \cdot r - 84
\end{aligned}$$

$$= c_{r-6} + 4 \cdot 6 \cdot r - 4 \cdot 21$$

$$= c_{r-6} + 4 \cdot \left(6 \cdot r - \sum_{i:=1}^{6} i\right)$$

$$=_{\substack{z := \#\text{rows} \\ \text{processed}}} c_{r-z} + 4 \cdot \left(z \cdot r - \sum_{i:=1}^{z} i\right)$$

$$=_{\substack{\text{Gauß formula} \\ \text{of summation}}} c_{r-z} + 4 \cdot [z \cdot r - z \cdot (z+1)/2]$$

$$= c_{r-z} + 4 \cdot z \cdot \frac{2 \cdot r - (z+1)}{2}$$

$$= c_{r-z} + 2 \cdot z \cdot (2 \cdot r - z - 1)$$

$$\vdots$$

$$:= c_{r-r} + 2 \cdot r \cdot (2 \cdot r - r - 1)$$

$$= c_0 + 2 \cdot r \cdot (r - 1)$$

$$:= 0 + 2 \cdot r \cdot (r - 1)$$

$$= 2 \cdot r \cdot (r - 1);$$

– Forward substitution:

$$c_1 := c_0 + 4 \cdot 0 := 0 + 0 = 0 := 4 \cdot 0$$

$$c_2 := c_1 + 4 \cdot 1 := 0 + 4 = 4 = 4 \cdot 1$$

$$c_3 := c_2 + 4 \cdot 2 := 4 + 8 = 12 = 4 \cdot 3$$

$$c_4 := c_3 + 4 \cdot 3 := 12 + 12 = 24 = 4 \cdot 6$$

$$c_5 := c_4 + 4 \cdot 4 := 24 + 16 = 40 = 4 \cdot 10$$

$$c_6 := c_5 + 4 \cdot 5 := 40 + 20 = 60 = 4 \cdot 15$$

$$:= 4 \cdot \sum_{i:=0}^{6-1} i$$

$$\vdots$$

$$c_r := 4 \cdot \sum_{i:=1}^{r-1} i$$

$$=_{\substack{\text{according to} \\ \text{C.-F. Gauß}}} \ddot\smile \ 4 \cdot \frac{(r-1) \cdot r}{2}$$

$$= 2 \cdot r \cdot (r - 1)$$

Statement: $c_r = 2 \cdot r \cdot (r - 1)$.

Proof: Induction on $r$ [$= \sqrt{n_r}$]:

– Basis: $r_0 := 0$

  - Principle: $c_0 = 0$ [no (computer-)row/column, no connection(s)]
  - Formula: $c_0 = 2 \cdot 0 \cdot (\ldots) = 0 \hat{=}$ Principle.
- Hypothesis: $c_{(r-1)} = 2 \cdot (r-1) \cdot [(r-1)-1]$.
- Step: $(r_0 \leq) r - 1 \to r\ (> r_0)$
  - Principle: $c_r := c_{r-1} + 4 \cdot (r-1)$

$$\begin{aligned} &\overset{!}{=} 2 \cdot (r-1) \cdot [(r-1)-1] + 4 \cdot (r-1) \\ &= 2 \cdot (r-1) \cdot [(r-2)+2] \\ &= 2 \cdot (r-1) \cdot r \end{aligned}$$

  - Formula: $c_r = 2 \cdot r \cdot (r-1) \hat{=}$ Principle.

Result: $c_r = 2 \cdot (r^2 - r) \;=\; 2 \cdot (n_r - \sqrt{n_r}) =: c_{n_r}$.

Application (of the # connections related to both parameters $r$ and $n_r$):

Regarding the design of a grid architecture, the functional dependence between the # rows (= # columns), respectively $n_r$ := # computers, and the need for the # connections $c_r$ (resp. $c_{n_r}$) is interesting, which expresses itself regarding the parameter $r$ quadratically and related to the parameter $n_r$ linearly. Once the network is in use, in the course of time the net admin her/himself typically poses the question of the additional costs needed in the case of a potential enlargement: How many connections do we have to add when the current network employs a further row and column of computer nodes? Regarding the principal order of magnitude we need a linear number of additional connections related to the number of rows (= # columns), which corresponds to a square-root function related to the # computers to obtain the additional # connections.

Example:

Input value: $r-1 := 5$

Output version: directly via the closed formula:

$$c_{r-1} = 2 \cdot (r-1) \cdot ((r-1)-1) = 2 \cdot 5 \cdot (5-1) = 10 \cdot 4 = 40 = c_5$$

Input value: $n_{r-1} := 5^2 = 25$

Output version: directly via the closed formula:

$$c_{n_{r-1}} = 2 \cdot (n_{r-1} - \sqrt{n_{r-1}}) \underset{:=5}{\overset{r-1}{=}} 2 \cdot (25 - \sqrt{5^2}) = 2 \cdot 20 = 40 = c_{n_5}$$

Input values: $c_{r-1}$ and $r := (r-1) + 1 := (6-1) + 1 = 6$

Output version: recursively via the recurrence relation:

$$c_r = c_{r-1} + \underline{4 \cdot (r-1)} = c_{6-1} + 4 \cdot (6-1) = c_5 + 4 \cdot 5 = 40 + 20 = 60 = c_6$$

We illustrated the linear addition; the absolute # cables is also obtainable directly:

$$c_6 = 2 \cdot 6 \cdot (6 - 1) = 60.$$

Input values: $c(n_{r-1})$ and $r := (r - 1) + 1 := (6 - 1) + 1 = 6$

Output version: recursively via the recurrence relation:

$$c(n_r) = c(n_{r-1}) + \underline{4 \cdot \sqrt{n_{r-1}}} = c(n_{6-1}) + 4 \cdot \sqrt{n_{6-1}} = c(n_5) + 4 \cdot \sqrt{n_5} = 40 + 4 \cdot 5 = 60 = c(n_6)$$

This is the square-root type of addition; the absolute # cables is also obtainable directly:

$$c_{n_6} = 2 \cdot (n_6 - \sqrt{n_6}) = 2 \cdot (6^2 - 6) = 60.$$

**$6^{th}$ illustration: Increase of nodes in a quadratic grid architecture**

We stay on this pattern of connections of the two illustrations just presented. Starting with $r$ rows (and also columns), we have $r^2 =: n_r$ as the # corresponding nodes, where we now interpret these nodes as pixels. We already know the quadratic relationship between $n_r$ (# nodes) and the row (and column) parameter $r$ (= $\sqrt{n_r}$). Here, we are interested in the recurrence principle itself—how many nodes get added to an existing pattern of pixels still consisting of $r-1$ rows (columns, resp.) by enlarging with 1 additional row and column. (This is realizable[20] quicker as the procedure shown now in the following; however, it is interesting to see how the recurrence appears to again serve as a helpful tool.)

From a principle point of view, it seems that, at first glance, we obtain by each new row with the label $r$ (> 0) and by this new column with the identical label $r$ each time so many further nodes ($2 \cdot r$); in order not to count the common new corner node twice, we must still subtract 1, hence: $n_r := n_{r-1} + 2 \cdot r - 1$. We now test this idea regarding the difference of two node numbers, where the number of rows (and columns) differ by 1, via the aforementioned recurrence principle—and thereby check whether the quadratic structure remains. (This is not so easy to perform it; however, I just like to deliver the recurrence relation again as a "nice" device.)

Basic value: $n_0 := 0;$
Principle: $n_r := n_{r-1} + 2 \cdot r - 1.$

Why? Each dimension yields $r$ nodes in the first step, however, the corner node not twice.

We now show that both substitution versions yield the known formula:

– Backward substitution:

$$\begin{aligned} n_r &:= n_{r-1} + 2 \cdot r - 1 \\ &:= n_{r-2} + 2 \cdot (r - 1) - 1 + 2 \cdot r - 1 = n_{r-2} + 4 \cdot r - 4 \end{aligned}$$

20 $n_r = n_{r-1} + x \Longleftrightarrow x = n_r - n_{r-1} = r^2 - (r - 1)^2 = r^2 - (r^2 - 2r + 1) = r^2 - r^2 + 2r - 1 = 2r - 1.$

$$
\begin{aligned}
&:= n_{r-3} + 2 \cdot (r-2) - 1 + 4 \cdot r - 4 = n_{r-3} + 6 \cdot r - 9 \\
&:= n_{r-4} + 2 \cdot (r-3) - 1 + 6 \cdot r - 9 = n_{r-4} + 8 \cdot r - 16 \\
&:= n_{r-5} + 2 \cdot (r-4) - 1 + 8 \cdot r - 16 = n_{r-5} + 10 \cdot r - 25 \\
&:=^{z := \#\text{rows}}_{\text{processed}} n_{r-z} + (2 \cdot z) \cdot r - z^2 \\
&\vdots \\
&:= n_{r-r} + 2 \cdot r^2 - r^2 \\
&= n_0 + r^2 \\
&= 0 + r^2 \\
&= r^2;
\end{aligned}
$$

– Forward substitution:

$$
\begin{aligned}
n_1 &:= n_0 + 2 \cdot 1 - 1 = 0 + 1 = 1 \\
n_2 &:= n_1 + 2 \cdot 2 - 1 = 1 + 3 = 4 \\
n_3 &:= n_2 + 2 \cdot 3 - 1 = 4 + 5 = 9 \\
n_4 &:= n_3 + 2 \cdot 4 - 1 = 9 + 7 = 16 \\
n_5 &:= n_4 + 2 \cdot 5 - 1 = 16 + 9 = 25 \\
&\vdots \\
n_r &= r^2
\end{aligned}
$$

Statement: $n_r = r^2$.

Proof: Induction on $r$ [$= \sqrt{n_r}$]:
- Basis: $r_0 := 0$
  - Principle: $n_0 = 0$ [no rows, now (pixel-)nodes present]
  - Formula: $n_0 = 0^2 = 0 \mathrel{\hat{=}}$ Principle.
- Hypothesis: $n_{(r-1)} = (r-1)^2$.
- Step: $(r_0 \leq) \, r - 1 \to r \, (> r_0)$
  - Principle: $n_r := n_{r-1} + 2 \cdot r - 1$

$$
\begin{aligned}
&\overset{!}{=} (r-1)^2 + 2 \cdot r - 1 \\
&= r^2 - 2 \cdot r + 1 + 2 \cdot r - 1 \\
&= r^2
\end{aligned}
$$

  - Formula: $n_r = r^2 \mathrel{\hat{=}}$ Principle.

Result: Indeed, we have found the correct recurrence in order to enlarge the (quadratic) grid.

Example: Computer/Mobile/TV display.

We start by a *c*urrent *r*ow (and column) number $r_c :=_{\text{here}} 1.024$, by which we employ a (1.024 × 1.024)-pixel matrix.

Question: How many *f*urther pixel do we gain once we *e*nlarge the number of rows (columns) by 1?

Answer: Let $r_e := r_c + 1, f := n_{r_e} - n_{r_c}$.

The recurrence principle $n_{r_e} := n_{r_c} + f$ yields us the # further pixel:

$$\begin{aligned} f &:= n_{r_e} - n_{r_c} \\ &:= n_{(r_c+1)} - n_{r_c} \\ &:= n_{r_c} + 2 \cdot r_e - 1 - n_{r_c} \\ &= 2 \cdot r_e - 1 \\ &:= 2 \cdot (r_c + 1) - 1 \\ &= 2 \cdot (1.024 + 1) - 1 = 2.050 - 1 = 2.049. \end{aligned}$$

Naturally, the following is true: $1.024^2 + 2.049 = 1.025^2$ (not a footnote ☺).

This view on the recurrence principle is slightly unusual; I've just offered it additionally. Here, the main point of the recurrence was the incremental structure itself, the next step, relative to the present situation; the incremental *part* $p_{r_e} := 2 \cdot r_e - 1$ was linear.

Clearly, we usually strive for an absolute functional dependence, which here is of a quadratic nature: $r^2 =: n_r$.

### 7th illustration: # bijections with *n* elements

By $b_n$, we count the number of different *b*ijections starting from an $n$-elementary domain into a codomain with obviously the same cardinality.

Basic value: $b_1 := 1$;
Principle: $b_n := b_{(n-1)} \cdot n$.

Why? A bijection means that not just each domain element gets assigned to exactly 1 co-domain element; we also must ensure that different inputs yield different outputs (injection), and that all codomain elements are needed (surjection). Due to the fact that both sets (co/domain) are of equal size, such a bijective function reflects 1 specific order of selecting the codomain elements. Each of such orderings ("permutations") represents a different mapping. We treat now this counting number of different possibilities to choose the individual codomain elements. Let us first build the recursion principle: Based on $b_{n-1}$ permutations related to an $(n-1)$-elementary set, the new ($n^{\text{th}}$) element has $n$ different possibilities to choose exactly 1 out of $n$ target elements—independent of the different permutations[21] among the already existing $n-1$ elements. Based on the

21 See Subsection 5.4.1, where the "factorial" sign "!" gets described.

insight that this is of a multiplicative nature, we obtain, for $n > 1$, the construction principle stated above.

We now show that both substitution versions yield the known formula:

- Backward substitution:

$$\begin{aligned} b_n &:= b_{(n-1)} \cdot n \\ &:= b_{(n-2)} \cdot (n-1) \cdot n \\ &:= b_{(n-3)} \cdot (n-2) \cdot (n-1) \cdot n \\ &\vdots \\ &:= b_{(n-(n-1))} \cdot (n-(n-2)) \cdot \dots \cdot (n-2) \cdot (n-1) \cdot n \\ &= b_1 \cdot \prod_{i:=2}^{n} i \\ &:= 1 \cdot \prod_{i:=2}^{n} i \\ &= \prod_{i:=1}^{n} i \\ &=: n!; \end{aligned}$$

- Forward substitution:

$$\begin{aligned} b_2 &:= b_1 \cdot 2 := 1 \cdot 2 = 2 \\ b_3 &:= b_2 \cdot 3 := 2 \cdot 3 = 6 \\ b_4 &:= b_3 \cdot 4 := 6 \cdot 4 = 24 \\ &\vdots \\ b_n &= \prod_{i:=1}^{n-1} i \cdot n \\ &= \prod_{i:=1}^{n} i \\ &=: n!. \end{aligned}$$

Statement: $b_n = n!$.

Proof: Induction on $n$:

- Basis: $n_0 := 1$
  - Principle: $b_1 = 1$ [the single element can get selected just $1 \times$]
  - Formula: $b_1 = 1! = 1 \mathrel{\hat{=}}$ Principle.
- Hypothesis: $b_{(n-1)} = (n-1)!$.

- Step: $(n_0 \leq) n-1 \to n (> n_0)$
  - Principle: $b_n := n \cdot b_{n-1}$
    $$\stackrel{!}{=} n \cdot (n-1)!$$
    $$= n!$$
  - Formula: $b_n = n! \mathrel{\hat{=}} \text{Principle}.$

## 5.4 Orderings and selections

### 5.4.1 Permutations

We now treat the number of different orderings of $n$ objects; in the case they are all different, we can start our considerations as follows: Object 1 naturally has just 1 order. Object 2 can get placed in front of the element already there or behind it: $1 \cdot 2 = 2$. Object 3 could find its place in front of the currently first element, in front of the second one or behind of this currently final one, independent of the 2 possibilities to order these two objects already there; hence $2 \cdot 3 = 6$. Object 4 could get the place in front of the first, second, third or behind the last one, independent of the 6 possibilities to order these other objects already there, hence $6 \cdot 4 = 24$, etc. The result searched for is obtained by the incremental product:

$$1 \cdot 2 \cdot 3 \cdot \dots \cdot n =: \prod_{i:=1}^{n} i =: n!$$

—called "$n$ factorial". (A simple induction proof would easily confirm it.) It holds $0! = 1$, because $(n-1)! = n!/n_{[>0]}$. The next 7 cases should be easily remembered, at least via $n! := (n-1)! \cdot n_{[>0]}: \dots, 7! = 5040$. The factorial quickly produces very large numbers; this function even grows exponentially, which is easy to realize (including proof).[22] For example, $13! = 6{,}227{,}020{,}800$ and $69! > 10^{98}$ (again not a footnote ☺).

What happens, regarding the number of different orderings, once we produce the permutations just of a subset of $k$ objects? This number is called the *permutation coefficient*:

$$P(n,k) := \frac{n!}{(n-k)!},$$

because we do not care about the different placements of the remaining $n-k$ objects.

A further name for this counting solution is *falling factorial*, $n$ to the $k$ falling:

$$n^{\underline{k}} := \frac{n!}{(n-k)!} = \frac{(n-k)! \cdot \prod_{i:=1}^{k}(n-k+i)}{(n-k)!} = \prod_{i:=0}^{k-1}(n-i).$$

22 $n! = \prod_{i:=1}^{n} i >_{[n \geq n_0 := 4]} \prod_{i:=1}^{n} 2 = 2^n$; the factor to the left grows (linearly), the one to the right not.

The following four examples might illustrate how $P(n, k)$ contributes to the solution of various tasks:

1. In an organization with $n_{[\geq 3]}$ members, we elect three special ones to act in different positions: let us say president, vice-president and secretary; afterwards, the organization itself determines the final placement of the three selected members.

   Question: How many possibilities are there to form the leader trio?
   Answer: $P(n, 3)$ $[= (n-2) \cdot (n-1) \cdot n]$.
2. $k$ students like to take a seat in an exam room with $n$ stools.

   Question: How many different placements are possible?
   Answer: $P(n, k)$.
3. We treat a simple construction of a user code consisting of 2 different parts: for the prefix, we take $k_1$ different digits out of an $n_1$-elementary set, and for the suffix we take $k_2$ different characters out of an $n_2$ elementary set.

   Question: How many different user codes are possible?
   Answer: $P(n_1, k_1) \cdot P(n_2, k_2)$.
4. Given are two finite sets with their cardinalities $|D| =: k$ and $|C| =: n$.

   Question: How many injections $\ddot{\smile}$ from $D$ to $C$ are possible?
   Answer: $P(n, k)$.
   Illustration:[23]
   Starting with $k$ elements in the *d*omain $D$, we select $k$ objects of the *c*odomain $C$, therefore we have $\binom{n}{k}$ possibilities.[24] Once we assign in each such configuration these $k$ objects selected ($\in C$) with a placement number $(1, \ldots, k)$ and virtually permutate them in different orders (which corresponds to different functions), we would have $k!$ possibilities for that, which results in[25]:

   $$\binom{n}{k} \cdot k! = \frac{n!}{(n-k)! \cdot k!} \cdot k! = \frac{n!}{(n-k)!} = P(n, k).$$

   The final fraction notation can get explained immediately: The nominator is obvious: $n!$ potential orderings of the $n$ elements in $C$. Due to the fact that just $k$ of them are needed in a specific function, the other $n-k$ elements must get ignored—including all permutations, i. e., 1 representative out of these $(n-k)!$ nonpossibilities, is sufficient; this idea is performed by the denominator. (Professionally, this is called “dividing by the cardinality of the size of the equivalence class”). By doing so, we end up by the permutation coefficient $P(n, k)$ $[= n^{\underline{k}}]$.

---

23 The other three examples mentioned above are obvious; here, I just explain the # injective functions.

24 The vertical expression with the long parentheses (“Binomial coefficient”) gets introduced in detail in Subsection 5.4.2 (starting on p. 72); it means how often we can differently choose $k$ elements out of $n$.

25 See footnote 24 just presented.

Let us still test this $P(n,k)$ regarding its special case $P(n,n)$, the number of different orders of all $n$ [=: $k$] objects—yielding the # bijections.

Question: Does $P(n,n) = n!$?

Answer: Yes!

Illustration:

$$P(n,n) = n^{\underline{n}} = \frac{n!}{(n-n)!} = n!.$$

$$P(n,n) = \prod_{i:=1}^{n}(n-n+i) = \prod_{i:=1}^{n} i = n!.$$

$$P(n,n) = \prod_{i:=0}^{n-1}(n-i) = n \cdot (n-1) \cdot (n-2) \cdot \dots \cdot 1 = n!.$$

Let us now come to the *rising factorial* of $k$ to $n$:

$$n^{\overline{k}} := \prod_{i:=0}^{k-1}(n+i).$$

Here, we test just the special case $1^{\overline{n}}$, the product of all 1-successors up to $n$.

Question: Does $1^{\overline{n}} = n!$?

Answer: Yes!

Illustration:

$$1^{\overline{n}} := \prod_{i:=0}^{n-1}(1+i) = \prod_{i:=1}^{n} i =: n! \quad [= n^{\underline{n}}].$$

Interesting to know, additionally, is the solution of the following scenario:

How many visually different *o*rderings among $n$ objects do we have once some of them are identical—or even several different groups of identical objects exist?

We would have $k_1$ objects of type 1, $k_2$ of type 2, $k_3$ of type 3, etc., $k_j$ of type $j$; $\sum_{i:=1}^{j} k_i =: n$. Then the following idea yields the desired solution:

$$o_{n,(k_1,\dots,k_j)} := \frac{(\sum_{i:=1}^{j} k_i)!}{k_1! \cdot \dots \cdot k_j!} = \frac{n!}{\prod_{i:=1}^{j}(k_i!)}.$$

The explanation of the fraction sounds similar to the one in the application example 4 for $P(n,k)$ on page 70 regarding the # injective mappings; here, just 1 representative per type $i$ ($1 \le i \le j$) suffices as representative of the corresponding $k_i!$ nondifferent permutations, the reason why we divide $n!$ by $k_i!$ in each of the $j$ cases.

The following three examples apply this formula:

1.

$$o_{3,(2,1)} = \frac{3!}{2! \cdot 1!} = \frac{2! \cdot 3}{2! \cdot 1} = 3; \quad \texttt{Illustration:}$$
$$|\{(true, true, false), (true, false, true), (false, true, true)\}| = 3.$$

2.

$$o_{4,(2,1,1)} = \frac{4!}{2! \cdot 1!^2} = \frac{2! \cdot 3 \cdot 4}{2! \cdot 1} = 12; \quad \texttt{Illustration:}$$
$$|\{AABC, AACB, ABAC, ACAB, ABCA, ACBA,$$
$$BAAC, CAAB, BACA, CABA, BCAA, CBAA\}| = 12.$$

3.

$$o_{5,(2,3)} = \frac{5!}{2! \cdot 3!} = \frac{3! \cdot 4 \cdot 5}{3! \cdot 2 \cdot 1} = 10; \quad \texttt{illustration:}$$
$$|\{(b,b,g,g,g), (b,g,b,g,g), (b,g,g,b,g), (b,g,g,g,b), (g,b,b,g,g),$$
$$(g,b,g,b,g), (g,b,g,g,b), (g,g,b,b,g), (g,g,b,g,b), (g,g,g,b,b)\}| = 10.$$

Also here, let us finally test the special case that all $n$ are different:

Question: Does $a_{n,(k_1,\ldots,k_n)} = n!$ ?

Answer: Yes!

Illustration:

$$o_{n,(k_1,\ldots,k_n)} = \frac{n!}{\prod_{i:=1}^{n}(k_i!)} = \frac{n!}{1!^n} = n!\,.$$

### 5.4.2 Combinations

Now, we just deal with the number of different selections of $k$ out of $n$ objects, i. e., how many $k$-elementary subsets out of an $n$-elementary base set exist, which gets represented by the *Combinatorial coefficient* $C(n,k)$. The more popular notion is "Binomial coefficient" pronounced "$n$ choose $k$", as visualized in the following mathematical expression:

$$\binom{n}{k} := \frac{n!}{k! \cdot (n-k)!} \quad \left[ = \begin{cases} 0; & n < k \quad \texttt{(defined)} \\ 1; & (n = k) \lor (n \geq k = 0) \end{cases} \right] .$$

The difference, respectively relationship, to $P(n,k)$ is obvious: We do not consider the permutations of the selected $k$ objects, the reason why again just 1 representative of these $k!$ potential orderings suffices, which gets obtained by dividing the permutation coefficient by these $k!$:

$$C(n,k) = \frac{P(n,k)}{k!}.$$

The following idea simplifies the computation and keeps the memory requirement small by not producing intermediate results, which later on do not play any role; therefore, we define $m := \mathtt{max}(k, n-k)$ and notate it like this:

$$\binom{n}{k} = \frac{m! \cdot (m+1) \cdot (m+2) \cdot \cdots \cdot n}{m! \cdot \mathtt{min}(k, n-k)!} = \frac{\prod_{i:=1}^{n-m}(m+i)}{\mathtt{min}(k, n-k)!}.$$

The following four examples apply this formula:

1.

$$\binom{4}{2} = \left[\frac{\mathtt{max}(2, 4-2)! \cdot (2+1) \cdot (2+(4-2))}{\mathtt{max}(2,2)! \cdot \mathtt{min}(2,2)!} = \right]$$
$$\frac{\prod_{i:=1}^{4-2}(2+i)}{2!} = \frac{3 \cdot 4}{2} = \frac{3 \cdot 2 \cdot 2}{2} = 6;$$
$$\text{instead of: } \binom{4}{2} = \frac{4!}{2! \cdot 2!} = \frac{24}{2 \cdot 2} = \frac{24}{4} = 6.$$

2.

$$\binom{7}{3} = \left[\frac{\mathtt{max}(3, 7-3)! \cdot (4+1) \cdot (4+2) \cdot (4+(7-4))}{\mathtt{max}(3,4)! \cdot \mathtt{min}(3,4)!} = \right]$$
$$\frac{\prod_{i:=1}^{7-4}(4+i)}{3!} = \frac{5 \cdot 6 \cdot 7}{6} = 35;$$
$$\text{instead of: } \binom{7}{3} = \frac{7!}{3! \cdot 4!} = \frac{5040}{6 \cdot 24} = \frac{5040}{144} = \cdots = 35.$$

3.

$$\binom{9}{4} = \left[\frac{\mathtt{max}(4, 9-4)! \cdot (5+1) \cdot (5+2) \cdot (5+3) \cdot (5+(9-5))}{\mathtt{max}(4,5)! \cdot \mathtt{min}(4,5)!} = \right]$$
$$\frac{\prod_{i:=1}^{9-5}(5+i)}{4!} = \frac{6 \cdot 7 \cdot 8 \cdot 9}{24} = \frac{(6 \cdot 4) \cdot 2 \cdot 7 \cdot 9}{24} = 126;$$
$$\text{instead of: } \binom{9}{4} = \frac{9!}{4! \cdot 5!} = \frac{7! \cdot 8 \cdot 9}{24 \cdot 120} = \frac{5040 \cdot 8 \cdot 3 \cdot 3}{120 \cdot 24} = \cdots = 126.$$

4.

$$\binom{11}{3} = \left[\frac{\mathtt{max}(3, 11-3)! \cdot (8+1) \cdot (8+2) \cdot (8+(11-8))}{\mathtt{max}(3,8)! \cdot \mathtt{min}(3,8)!} = \right]$$
$$\frac{\prod_{i:=1}^{11-8}(8+i)}{3!} = \frac{9 \cdot 10 \cdot 11}{6} = \frac{3 \cdot 3 \cdot 2 \cdot 5 \cdot 11}{3 \cdot 2} = 165;$$
$$\text{instead of: } \binom{11}{3} = \frac{11!}{3! \cdot 8!} = \frac{11!}{6 \cdot 5040 \cdot 8} = \frac{11!}{6 \cdot 40320} = \cdots = 165.$$

The idea is to reduce the fraction right from the beginning; the figures, therefore, stay as small as possible.

We now come to a very obvious feature: The number of possibilities to choose $k$ elements out of an $n$-elementary set is identical to the # choosing $n-k$ elements: In the first case, we select $k$ and ignore $n-k$ "partner" elements and in the second case it is just the opposite consideration where we select $n-k$ and ignore $k$ "partner" elements; only the semantics[26] of the objects differs—to select them or not. (In any case, there is no further choice for the nonselected "partner" objects.) Due to the fact that these two cases are symmetrical to each other, we talk about the *Binomial symmetry*, with identical syntax[27]:

$$\binom{n}{k} = \binom{n}{n-k}.$$

This handy feature is provable *directly*:

$$\binom{n}{n-k} = \frac{n!}{(n-k)! \cdot (n-(n-k))!} = \frac{n!}{(n-k)! \cdot (n-n+k)!} = \frac{n!}{k! \cdot (n-k)!} = \binom{n}{k}.$$

"*Pascal*'s triangle" (perhaps still known from school) can be shown similarly (perhaps in a more elegant fashion possible, too):

$$\begin{aligned}
\binom{n}{k} &= \binom{n-1}{k} + \binom{n-1}{k-1} \\
&= \frac{(n-1)!}{k! \cdot (n-1-k)!} + \frac{(n-1)!}{(k-1)! \cdot (n-1-(k-1))!} \\
&= \frac{(n-1)!}{k! \cdot (n-k-1)!} + \frac{(n-1)!}{(k-1)! \cdot (n-1-k+1)!} \\
&= \frac{(n-1)!}{k! \cdot (n-k-1)!} \cdot \frac{n}{n} + \frac{k}{k} \cdot \frac{(n-1)!}{(k-1)! \cdot (n-k)!} \cdot \frac{n}{n} \\
&= \frac{n!}{k! \cdot (n-k-1)! \cdot n} + \frac{n! \cdot k}{k! \cdot (n-k)! \cdot n} \\
&= \frac{n!}{k!} \cdot \left( \frac{n-k}{(n-k-1)! \cdot (n-k) \cdot n} + \frac{k}{(n-k)! \cdot n} \right) \\
&= \frac{n!}{k!} \cdot \left( \frac{n-k}{(n-k)! \cdot n} + \frac{k}{(n-k)! \cdot n} \right) \\
&= \frac{n!}{k! \cdot (n-k)!} \cdot \frac{n-k+k}{n} \\
&= \frac{n!}{k! \cdot (n-k)!} \quad \ddot\smile .
\end{aligned}$$

So far, the binomial coefficient is computable on at least two ways: "closed"[28] and recursively[29]. Now we offer a further computation variant: successively[30], namely an incremental sum:[31]

26 Meaning.

27 Structure.

28 Here as quotient of two factorials.

29 Going back to previous subresults—as by the addition just presented.

30 Stepwisely.

31 In the following formula, the bottom sum index is the upper binomial parameter increasing by 1.

$$\binom{n}{k} = \sum_{[0\leq]\, i:=k-1}^{[0\leq]\, n-1} \binom{i}{k-1}.$$

Figure 5.4 likes to illustrate this.

| | **k** | 0 | 1 | 2 | 3 | 4 | … |
|---|---|---|---|---|---|---|---|
| **n** | | | | | | | |
| 0 | | 1 | 0 | 0 | 0 | 0 | … |
| 1 | | 1 | 1 | 0 | 0 | 0 | … |
| 2 | | 1 | 2 | 1 | 0 | 0 | … |
| 3 | | 1 | 3 | 3 | 1 | 0 | … |
| 4 | | 1 | 4 | 6 | 4 | 1 | … |
| ⋮ | | | | | | | ⋱ |

**Figure 5.4:** Binomial sum.

Finally, we shed some light to further interesting items.

The “central” Binomial coefficient is the one where the bottom parameter $k \approx n/2$. Once we visualize all possible $C(n, k)$ in a powerset-lattice with $n+1$ levels[32]—at the bottom $C(n, 0)$, on the higher-levels (“step-by-step”) the next larger $k$, etc., up to $C(n, n)$—we see that the number of elements at the level in the middle with the label (number) $\lfloor n/2 \rceil$ is maximal[33] $C(n, \lfloor n/2 \rceil) \geq C(n, k), \forall k$. If $n$ is even, we do not need a rounding function; in the case that $n$ is odd, the unusual “do not care” rounding symbol should signalize that it does not matter whether we round to the “floor” or “ceiling” (LaTeX wording) $[C(n, (n-1)/2) = C(n, (n+1)/2)]$, due to the known binomial symmetry:

$$\binom{n}{\lfloor \frac{n}{2} \rceil} = \begin{cases} \binom{n}{\frac{n}{2}}; & \text{even}(n) \\ \binom{n}{\frac{n-1}{2}}; & \text{odd}(n) \end{cases} = \begin{cases} \binom{n}{\frac{n}{2}}; & \text{even}(n) \\ \binom{n}{\frac{n+1}{2}}; & \text{odd}(n). \end{cases}$$

2 illustrations:

$$\binom{1}{\lfloor \frac{1}{2} \rceil} = \binom{1}{\frac{1-1}{2}} = \binom{1}{\frac{0}{2}} = \binom{1}{0}$$
$$= \binom{1}{\frac{1+1}{2}} = \binom{1}{\frac{2}{2}} = \binom{1}{1} \quad [= 1].$$

32 Labeled from 0 to $n$, the possible values of $k$ (see Figure 1.1 starting at p. 10 in Section 1.3).

33 The reason why the $C$ expression in the text above is called the *maximal* Binomial coefficient.

$$\binom{3}{\lfloor\frac{3}{2}\rfloor} = \binom{3}{\frac{3-1}{2}} = \binom{3}{\frac{2}{2}} = \binom{3}{1}$$
$$= \binom{3}{\frac{3+1}{2}} = \binom{3}{\frac{4}{2}} = \binom{3}{2} \quad [= 3].$$

Once $n$ is even, the central Binomial coefficient does exist only 1 ×; in the other case where $n$ is odd, this maximal Binomial coefficient appears twice; we have to keep this interesting feature always in mind in corresponding counting problems.

The Binomial coefficient represents the main part of the *Binomial Theorem*

$$(a+b)^n = \sum_{k:=0}^{n}\left(\binom{n}{k}\cdot a^{(n-k)}\cdot b^k\right).$$

With this knowledge, it is then easy to prove a statement, which in earlier math days, was controversial: $0^0 = 1\ [\neq 0]$:

$$1 = 1^0 = (0+1)^0 = \binom{0}{0}\cdot 0^{(0-0)}\cdot 1^0 = 1\cdot 0^0\cdot 1 = 0^0.$$

By the important specialization $a := b := 1$, we obtain the following:

$$(1+1)^n = \sum_{k:=0}^{n}\left(\binom{n}{k}\cdot 1^{(n-k)}\cdot 1^k\right) \iff \sum_{k:=0}^{n}\binom{n}{k} = 2^n.$$

Due to the fact that $C(n,k)$ provides us the number of $k$-elementary subsets of an $n$-elementary finite set, the above mentioned formula clarifies that the (power-)set of all ("improper") subsets has exactly $2^n$ elements—pointing to an exponential number.

Regarding the first 14 $n$ input cases, we should be able to remember this exponential output.[34]

The following five examples show how $C(n,k)$ contributes to the solution of certain tasks:

1. In an organization with $n$ members, we might choose $k$ board members.

   Question: How many selections are possible (in the case all of them are "ok" ☺)?
   Answer: $C(n,k)$.
2. Out of a squad consisting of $n$ players we have to select $k$ players.[35]

   Question: How many nomination sheets (independent of positions) are possible right ahead of the start of the game?
   Answer: $C(n,k)$.

34 $2^0 = 1, \ldots, 2^{10} = 1024, \ldots, 2^{13} = 8192$; $2^i :=_{[i>0]} 2^{(1)}\cdot 2^{(i-1)}$.

35 Typically, as in the team *FC Bayern München* ☺, there naturally exists "selection pressure": $k < n$; however, also for $k = n$ and $k > n$ the approach yields the correct answer.

3. We again consider the well-known ;) world of bit-streams.[36]

   Question: How many $n$-ary bit-streams with at least $k \times true$ are constructible?[37]
   Answer: $\sum_{i:=k}^{n} C(n, i)$.
4. Let us stay for a short while in this area important for Computer Science.

   Question: How many $n$-ary binary strings are possible when # *false*- = # *true*-bits?
   Answer:

$$\begin{cases} 0; & \text{odd}(n), \\ \binom{n}{\frac{n}{2}}; & \text{even}(n). \end{cases}$$

   In this example, we see the central Binomial coefficient in action.
5. Let us finally visit the mathematical background of game design. We are in the center of the coordinate system, 2d-like localized by "(0, 0)".

   Question: How many minimal axis-parallel paths from the origin to the point "$(a, b)$"[38] could get offered ($|a|$ steps horizontally, $|b|$ vertically)[39]?
   Answer: (see Figure 5.5):

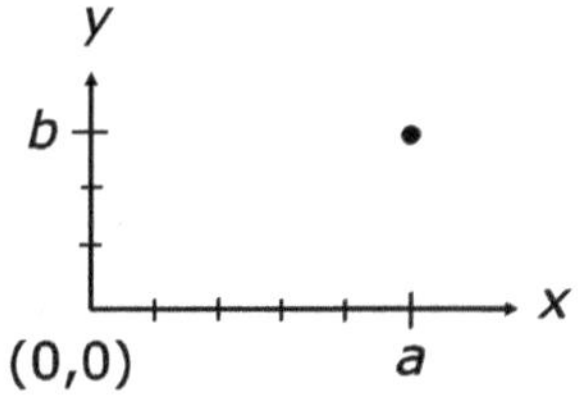


**Figure 5.5:** 2d-trajectory.

$$\binom{|a| + |b|}{|a|} =_{\substack{\text{Binomial}\\\text{symmetry}}} \binom{|a| + |b|}{(|a| + |b|) - |a|} = \binom{|a| + |b|}{|b|} = \binom{|b| + |a|}{|b|},$$

which would also reflect the answer for the symmetric[40] question regarding the number of different paths to the point "$(b, a)$": In all cases, the basis for our consideration is the total number of steps (traversing both in $x$- and $y$-direction) to go anyway, i. e., the sum of both absolute coordinate values, not the order of possible visits of the various pixels in each dimension, which in general explains the formulae

36 Strings consisting of *false* ("0") and/or *true* ("1").

37 Based on a fixed $n$—leading 0-values (*false*-bits) as prefixes are therefore allowed.

38 $\in \mathbb{Z}^2 \setminus (0, 0)$; independent of the signs—we have a symmetry regarding the axes $x$ and $y$.

39 $|n| :=_{\text{(here)}}$ absolute value of a number.

40 W. r. t. the angle bisector across the odd-numbered quadrants.

for the corresponding # paths with minimal length: We start our path (performing $|a| + |b|$ steps) in the origin of the coordinate system [(0,0)].

$1^{st}$ consideration: Let us regard the $y$-axis, to finally reach, step-by-step, position $b$; on this way, we receive $|a|$ times a wind deviation horizontally (in $x$-direction).[41] The number of different possibilities of such movements is $C(|a| + |b|, |a|)$.

$2^{nd}$ consideration: Let us regard the $x$-axis, to finally reach, step-by-step, position $a$; on this way, we feel $|b|$ times a tiny earthquake vertically (in $y$-direction).[42] The number of different possibilities of such movements is $C(|a| + |b|, |b|)$.

We further obtain the formula also by a completely different consideration: Subsection 5.4.1 (p. 71) allows us to count the # configurations of $n$ objects once several types of identical objects exist: The 2 dimensions ($x$- and $y$-direction) now represent the different types, whereas the identical feature is the corresponding direction of movements in a certain dimension;[43] by

$$k_1 := |a|, \quad k_2 := |b|, \quad n := k_1 + k_2 := |a| + |b|,$$

we then obtain for the number of different 2-dimensional trajectories:

$$\frac{(\sum_{i:=1}^{2} k_i)!}{\prod_{i:=1}^{2}(k_i!)} = \frac{(|a| + |b|)!}{|a|! \cdot |b|!} = \frac{(|a| + |b|)!}{|a|! \cdot ((|a| + |b|) - |a|)!} = \binom{|a| + |b|}{|a|}.$$

Even more complex scenarios with intermediate stations (important for delivery tasks) are demonstrated in my Computer Science book[44] cited in the bibliography, pp. 12–14.

## 5.5 Stirling and Bell numbers

### 5.5.1 Stirling numbers of the $1^{st}$ kind

Principally, this kind of numbers treat the coefficients of $x^k$ in the sum notation of the expression $\prod_{i:=0}^{n-1}(x - i) =: j_1$; they are named according to *James Stirling*: $s_1(n, k)$:

$$j_1 = \sum_{k:=0}^{n} (s_1(n, k) \cdot x^k).$$

41 $a < 0$ east-, $a > 0$ west-wind; $a = 0$ no wind ☺—in this special case, we immediately reach $b$, staying at the $y$-axis.

42 For a negative $b$, each vertical displacement moves us 1 pixel level lower—a usual case in 2d games.

43 The pixels get visited in subsequent order, starting in the origin of the coordinate system, by performing each time 1 step in the corresponding direction.

44 Initial edition; second edition in progress, still in German.

This recursion defines $s_1(n,k) :=$

$$\begin{cases} 0; & (n < k) \vee (n > k = 0), \\ 1; & n = k, \\ s_1(n-1,k-1) - (n-1) \cdot s_1(n-1,k); & n > k > 0. \end{cases}$$

Let $n := 3$, $M := \{0,\dots,n\}$; we obtain the following $|M|$ values for $s_1(n,k)$, $k \in M$: $(-)0, 2, -3, 1$.

Interesting for us is the interpretation as (*Stirling*) *cycle number* via $|s_1(n,k)| =:$

$$\begin{bmatrix} n \\ k \end{bmatrix}, \quad \texttt{to say (proposal)} : n \texttt{ cycle } k,$$

the number of so-called *cyclic partitions* of an $n$-elementary set in $k$ nonempty "cycles":

The task is to partition an $n$-set in $k$ nonempty parts, where we consider all possibilities of different ring formations[45] of the objects—in a sense to be placed around a round table; thereby, neither the specific place at the table nor a specific table counts;[46] see also the special case $|s_1(m,1)|$, which could have solved the challenge in the illustration of the *Rule of Quotient* in the Subsection 5.1.3 (starting at p. 48).[47]

Figure 5.6 illustrates the Stirling number of the $1^{\text{st}}$ kind for $k$ tables with at least 1 person:

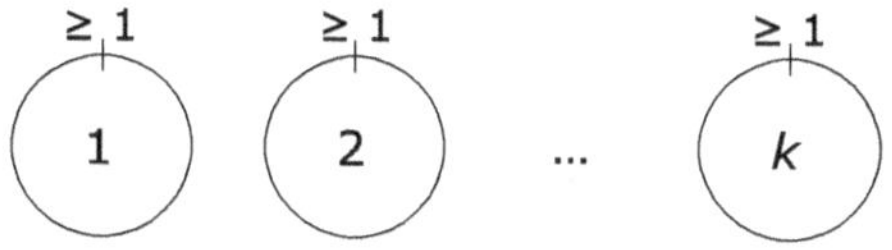


**Figure 5.6:** Stirling-1.

Example:

$$\begin{bmatrix} 4 \\ 2 \end{bmatrix} = |s_1(4,2)| =_{\texttt{recursively}} \cdots = |11| = 11$$

$$=_{\texttt{explicitly}} |\{\{(1),\ (2,3,4)\}, \{(1),\ (2,4,3)\}, \{(2),\ (1,3,4)\}, \{(2),\ (1,4,3)\}, \{(3),\ (1,2,4)\}, \{(3),\ (1,4,2)\}, \{(4),\ (1,2,3)\}, \{(4),\ (1,3,2)\}, \{(1,2),\ (3,4)\}, \{(1,3),\ (2,4)\}, \{(1,4),\ (2,3)\}\}|.$$

45 Related to "is positioned exactly 1 position left of another object" (or "right of", depending on the viewing direction).

46 It just depends on the relative ordering at the table according to the left|right neighboring rule.

47 There, all $n := m$ "objects" at $k := 1$ table: $(m-1)!$

Performing the computation for the various assignments of $k$ for a given $n$ stepwise, we regard that for each incrementation (by 1) of $k$ the sign of $s_1(n,k)$ changes.[48] Obviously, the cycle number can just be positive—the reason for the following relationship:

$$s_1(n,k) = (-1)^{(n-k)} \cdot \begin{bmatrix} n \\ k \end{bmatrix} \quad | : (-1)^{(n-k)} \qquad \Longleftrightarrow$$

$$\begin{bmatrix} n \\ k \end{bmatrix} = \frac{s_1(n,k)}{(-1)^{(n-k)}} = \begin{cases} -s_1(n,k); & \mathsf{odd}(n-k),\ s_1(n,k) \le 0, \\ +s_1(n,k); & \mathsf{even}(n-k),\ s_1(n,k) \ge 0. \end{cases}$$

(The cases are not disjoint; the only common ["= 0"-]case in both parts however yields the same result, see also the content of the parenthesis expression of footnote 48.)

$$= |s_1(n,k)| = |s_1(n-1,k-1) - (n-1) \cdot s_1(n-1,k)|$$

$$= \begin{cases} |-[{n-1 \atop k-1}] - (n-1) \cdot (+[{n-1 \atop k}])|; & s_1(n-1,k) \ge 0 \\ |+[{n-1 \atop k-1}] - (n-1) \cdot (-[{n-1 \atop k}])|; & s_1(n-1,k) \le 0 \end{cases}$$

$$= \begin{cases} |-([{n-1 \atop k-1}] + (n-1) \cdot [{n-1 \atop k}])|; & \mathsf{odd}(n-k) \\ |+([{n-1 \atop k-1}] + (n-1) \cdot [{n-1 \atop k}])|; & \mathsf{even}(n-k) \end{cases}$$

$$= \begin{bmatrix} n-1 \\ k-1 \end{bmatrix} + (n-1) \cdot \begin{bmatrix} n-1 \\ k \end{bmatrix}.$$

Statement:

$$z(n) := \sum_{k:=0}^{n} \begin{bmatrix} n \\ k \end{bmatrix} = n! \quad \left( = n \cdot (n-1)! = n \cdot \begin{bmatrix} n \\ 1 \end{bmatrix} \right).$$

Proof: Induction on $n$:

Start: $n_{0_{\mathtt{initially}}} := 0$

$$z_{\underline{\mathtt{P}}\mathtt{rinciple}}(0) := \sum_{k:=0}^{0} \begin{bmatrix} 0 \\ k \end{bmatrix} = \begin{bmatrix} 0 \\ 0 \end{bmatrix} = 1 = 0! = z_{\underline{\mathsf{F}}\mathsf{ormula}}(0).$$

In the induction step, still to come, it will be comfortable to have $n-1 \ge 1$.

Basis: $n_0 := 1$

$$z_{\mathsf{P}}(1) := \sum_{k:=0}^{1} \begin{bmatrix} 1 \\ k \end{bmatrix} = \begin{bmatrix} 1 \\ 0 \end{bmatrix} + \begin{bmatrix} 1 \\ 1 \end{bmatrix} = 0 + 1 = 1 = 1! = z_{\mathsf{F}}(1).$$

Hypothesis:

48 "Alternates" (0 considered as negative or positive number).

$$z(n-1) := \sum_{k:=0}^{n-1} \begin{bmatrix} n-1 \\ k \end{bmatrix} = (n-1)!.$$

Step: $(n_0 \leq)\, n-1 \to n\, (> n_0)$

$$\begin{aligned}
z_P(n) &:= \sum_{k:=0}^{n} \begin{bmatrix} n \\ k \end{bmatrix} = \begin{bmatrix} n \\ 0 \end{bmatrix} + \sum_{k:=1}^{n} \left( (n-1) \cdot \begin{bmatrix} n-1 \\ k \end{bmatrix} + \begin{bmatrix} n-1 \\ k-1 \end{bmatrix} \right) \\
&= 0 + (n-1) \cdot \sum_{k:=1}^{n} \begin{bmatrix} n-1 \\ k \end{bmatrix} + \sum_{k:=1}^{n} \begin{bmatrix} n-1 \\ k-1 \end{bmatrix} \\
&= (n-1) \cdot \left( \sum_{k:=1}^{n-1} \begin{bmatrix} n-1 \\ k \end{bmatrix} + \begin{bmatrix} n-1 \\ n \end{bmatrix} \right) + \sum_{k:=1}^{n-1} \begin{bmatrix} n-1 \\ k-1 \end{bmatrix} + \begin{bmatrix} n-1 \\ n-1 \end{bmatrix} \\
&= (n-1) \cdot \left( \sum_{k:=0}^{n-1} \begin{bmatrix} n-1 \\ k \end{bmatrix} - \begin{bmatrix} n-1 \\ 0 \end{bmatrix} + 0 \right) + \sum_{k:=0}^{n-1} \begin{bmatrix} n-1 \\ k \end{bmatrix} - \begin{bmatrix} n-1 \\ n-1 \end{bmatrix} + \begin{bmatrix} n-1 \\ n-1 \end{bmatrix} \\
&\overset{!}{=} (n-1) \cdot (n-1)! + (n-1)! = ((n-1)+1) \cdot (n-1)! = n \cdot (n-1)! \\
&= n! = z_F(n).
\end{aligned}$$

### 5.5.2 Stirling numbers of the 2nd kind

Historically, these numbers are the coefficients of $\prod_{i:=0}^{k-1}(x-i) =: j_2$ in $x^k$ notated as sum of products; we name them according to *James Stirling*: $s_2(n,k)$:

$$x^n = \sum_{k:=0}^{n} (s_2(n,k) \cdot j_2).$$

The following recursion defines $s_2(n,k) :=$

$$\begin{cases} 0; & (n < k) \vee (n > k = 0), \\ 1; & (n = k) \vee (n > k = 1), \\ s_2(n-1,k-1) + k \cdot s_2(n-1,k); & n > k > 1. \end{cases}$$

Let $M := \{0, \ldots, 5\}$; Figure 5.7 now shows us the $|M|^2$ values for $s_2(n,k)$ $[k, n \in M]$. As an incremental sum notation we obtain the equation $s_2(n,k) =$

$$\sum_{i:=k-1}^{n-1} \left( \binom{n-1}{i} \cdot s_2(i,k-1) \right).$$

Stirling numbers of the 2nd kind have a particular relationship with those of the 1st kind, as we see soon; due to the alternating sign of $s_1(n,k)$—in contrast to the per-

| k / n | 0 | 1 | 2 | 3 | 4 | 5 | ... |
|---|---|---|---|---|---|---|---|
| 0 | 1 | 0 | 0 | 0 | 0 | 0 | |
| 1 | 0 | 1 | 0 | 0 | 0 | 0 | |
| 2 | 0 | 1 | 1 | 0 | 0 | 0 | |
| 3 | 0 | 1 | 3 | 1 | 0 | 0 | |
| 4 | 0 | 1 | 7 | 6 | 1 | 0 | |
| 5 | 0 | 1 | 15 | 25 | 10 | 1 | |

**Figure 5.7:** Stirling-2.

manently positive one in $s_2(n,k)$—they neutralize themselves in the following sum of products:

$$\sum_{k:=0}^{n}(s_1(n,k)\cdot s_2(k,i)) = \sum_{k:=0}^{n}(s_2(n,k)\cdot s_1(k,i)) = \begin{cases}1; & n=i,\\ 0; & n\neq i.\end{cases}$$

Interesting for us is the interpretation as (*Stirling*)*subset number* via $s_2(n,k)$ =:

$$\left\{ {n \atop k} \right\}, \quad \texttt{to say (proposal) : } n \texttt{ partition } k,$$

the number of partitions of an $n$-elementary set in $k$ subsets: One divides an $n$-set in $k$ parts, without considering orderings. This automatically means:

$$\left\{ {n \atop k} \right\} \leq \left[ {n \atop k} \right].$$

For $k \in \{n-1, n\}$, the following equation holds:

$$\left\{ {n \atop n-1} \right\} =_{n>0} \left[ {n \atop n-1} \right] =_{n>1} \left[ {n-2 \atop n-2} \right] \cdot \binom{n}{2} = n\cdot(n-1)/2:$$

At first, we create $n-1$ subsets, therefore we have $n-2$ singletons (subsets with exactly just 1 element, without any cycle variants) and the remaining 2 elements are in the final 1 subset (again without any further cycle variant); for the selection just mentioned, we have $C(n,2)$ possibilities, and for the partition mentioned first we do not have any further choice:[49] $|s_1(n-2,n-2)| = 1$.

$$\left\{ {n \atop n} \right\} = \left[ {n \atop n} \right] = 1 =_{n>0} \left\{ {n \atop 1} \right\}:$$

49 $\exists!\,1$ possibility placing $n-2$ elements in $n-2$ cycles of "length" (:= # elements) 1.

The first pattern requires $n$ subsets, such that in each one just 1 element is inside, the reason why the pattern in the middle cannot produce more cycles,[50] which happens exactly 1×—as in the latter case producing just 1 nonempty subset, in which 1× all elements are contained.

The following example, in the area of functions, shows an application of $s_2(n,k)$:

Question: How many surjective $f$unctions are possible from an $n$- into a $k$-set?

Answer: Let $n := |D|, k := |C|; f : D \to C$.

Then we have $s_2(n,k) \cdot k!$ surjections.

Illustration: All $k$ elements of $C$ must get selected, therefore, the $n$ elements of $D$ have to get clustered in $k$ groups[51]—and for the realization of all different orderings of $k$ groups we have $k!$ permutations.

Let us check this application in the special case of $k = n$, in order to get the confirmation by the formula of the computation of the # bijections: $n! \cdot s_2(n,n) = n!$—ok.[52]

As one of the highlights of this chapter, I now present the derivation of the closed formula in the special case $s_2(n,2)$ for an arbitrary $n_{[>1]}$; this formula yields the number of possibilities to divide a given nonempty set of $n$ elements in 2 nonempty parts:

$$\left\{ {n \atop 2} \right\} = \sum_{i:=1}^{\lfloor \frac{n}{2} \rfloor} \binom{n}{i} - \begin{cases} \frac{\binom{n}{\frac{n}{2}}}{2}; & \text{even}(n) \\ 0; & \text{odd}(n) \end{cases} = \frac{\sum_{i:=0}^{n} \binom{n}{i}}{2} - \binom{n}{0} = \frac{2^n}{2^1} - 1 = 2^{(n-1)} - 1.$$

Let us first inspect the subtraction with the case analysis:

We start by 1-elementary subsets and put the other elements in the corresponding ($n$–1)-elementary subsets (objects), in Figure 5.8 visualized by dotted lines; in the following: the cardinality of the small-elementary subsets gets enlarged step-by-step up to the half (of $n$, the highest number in the subset lattice[53]).

Proceeding further on is nonsense, because the level objects of the higher second half of the lattice already serve as partners of subsets in the lower first half. For odd-numbered $n$, we are done.[54] For even-numbered $n$, we should remember that the partners of the subsets in the middle are placed on the same level, i. e., on this widest level[55] $n/2$ in the powerset lattice we must consider just 1 half of the central binomial coefficient, because the other half already presents the partner subsets there; see the broken lines in

50 Of length 1.

51 $n > k$ allowed—injectivity is not required.

52 See also the special case $P(n,n)$ in Subsection 5.4.1 (p. 71).

53 See Section 1.3 (starting at p. 10).

54 Is $n$ odd, the powerset-lattice has an even number $(n+1)$ of levels (labeled $0 \ldots n$).

55 There, # objects is maximal: the "maximal" binomial coefficient $C(n, n/2) \geq C(n,i)\ \forall i$.

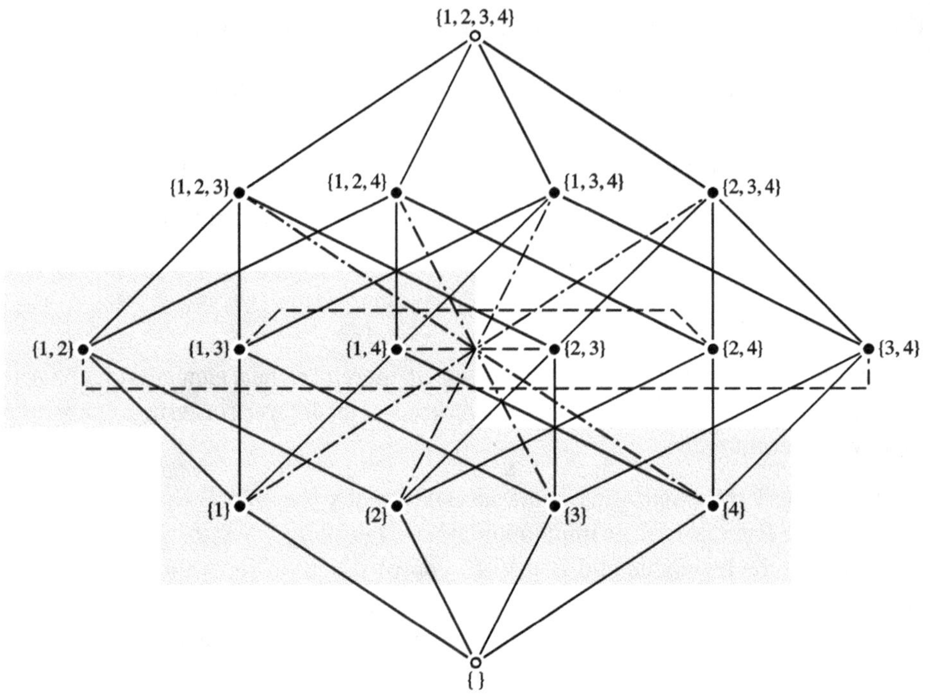


**Figure 5.8:** Half-way split.

the picture.[56] Due to the fact that the subsets of this middle level are already completely counted by the top sum index, we must finally subtract half of these $C(n, n/2)$ subsets. (For $n$ being odd, rounding up the upper sum index would end up incorrectly, because the "middle" level, which appears twice[57], would have got counted $1\times$ too often.)

Now we explain the subtraction of the binomial coefficient $C(n, 0)$ from the fraction mentioned above.

The formula just discussed corresponds—regardless of the parity[58] of $n$—to the difference[59] of half of the total sum and the binomial coefficient $C(n, 0)$: The number of nodes in the lower half of the powerset-lattice equals, due to the binomial symmetry, exactly the half of the total number;[60] due to the fact that we consider only nonempty subsets, we delete the empty case, the reason why we subtract $C(n, 0)$ [= 1] at the end.

56 We only count the number of different partitions of $n$ elements in 2 parts—here halfs.

57 Binomial symmetry: $C(n, \lfloor n/2 \rfloor) = C(n, n - \lfloor n/2 \rfloor) = C(n, \lceil n/2 \rceil)$ [the partners of the $C(n, \lfloor n/2 \rfloor)$].

58 "Even" or "odd".

59 Which follows there.

60 In the "even" case, the fraction line produces the halving of the powerset-lattice through the middle of the central binomial coefficient of the level $n/2$ (as in Figure 5.8), in the "odd" case the division between the two "middle" levels $\lfloor n/2 \rfloor$ and $\lceil n/2 \rceil$ with the identical common number of partner subsets.

Now to the closed form:

For the sum in the numerator, we already have the expression $2^n$ at hand, from which we, of course, only take its half, and finally subtract the 1 nonpossibility—done.

Illustration:

$$\begin{aligned} s_2(4,2) &= 2^{(4-1)} - 1 = 2^3 - 1 = 8 - 1 = 7 \\ &=_{\text{e. g.}} \Big|\big\{\{\{1\},\{2,3,4\}\},\{\{2\},\{1,3,4\}\},\{\{3\},\{1,2,4\}\},\{\{4\},\{1,2,3\}\}, \\ &\qquad \{\{1,2\},\{3,4\}\},\{\{1,3\},\{2,4\}\},\{\{1,4\},\{2,3\}\}\big\}\Big| \leq 11 = |s_1(4,2)|. \end{aligned}$$

### 5.5.3 Bell numbers

Let us finally collect all possible partitions of a given set of cardinality $n$ [$\geq 0$], independent of the number of resulting parts. By considering all cases of $k$ [$\in \{0,\ldots,n\}$] in $s_2(n,k)$, we obtain via summation of all Stirling partitions the so-called *B*ell number:

$$B(n) = \sum_{k:=0}^{n} s_2(n,k) = \sum_{k:=0}^{n} \begin{Bmatrix} n \\ k \end{Bmatrix}.$$

The following recursion defines $B(n) :=$

$$\begin{cases} 1; & n \leq 1, \\ \sum_{i:=0}^{n-1}(\binom{n-1}{i} \cdot B(i)); & n \geq 1. \end{cases}$$

This nondisjoint case analysis allows for its just 1 common case ($n = 1$) to yield the function value also via replication of the recursion base ($n := 0$) immediately: $B(1)$ [$:= B(0)$] $:= 1$.

Let us consider some concrete values:

$$\begin{aligned} B(0) &= 1, \\ B(1) &= \sum_{i:=0}^{1-1}\left(\binom{0}{i} \cdot B(i)\right) = 1 \cdot B(0) = 1, \\ B(2) &= \sum_{i:=0}^{2-1}\left(\binom{1}{i} \cdot B(i)\right) = 1 \cdot B(0) + 1 \cdot B(1) = 2, \\ B(3) &= \sum_{i:=0}^{3-1}\left(\binom{2}{i} \cdot B(i)\right) = 1 \cdot B(0) + 2 \cdot B(1) + 1 \cdot B(2) = 5 \\ &=_{\text{e. g.}} |\{\{\{a,b,c\}\},\{\{a\},\{b,c\}\},\{\{b\},\{a,c\}\},\{\{c\},\{a,b\}\},\{\{a\},\{b\},\{c\}\}\}|, \\ B(4) &= \sum_{i:=0}^{4-1}\left(\binom{3}{i} \cdot B(i)\right) = 1 \cdot B(0) + 3 \cdot B(1) + 3 \cdot B(2) + 1 \cdot B(3) \\ &= 1 \cdot 1 + 3 \cdot 1 + 3 \cdot 2 + 1 \cdot 5 = 15 \end{aligned}$$

$$
\begin{aligned}
&=_{5.9}^{\text{Figure}} B(3) + ((B(1) + B(2)) + (B(2) + B(3))) \\
&= 1 \cdot B(1) + (1+1) \cdot B(2) + (1+1) \cdot B(3) \\
&= 1 \cdot B(0) + (2 \cdot B(1) + 1 \cdot B(2)) + ((1 \cdot B(2) + (1 \cdot B(1) + 1 \cdot B(2))) + 1 \cdot B(3)) \\
&= \binom{3}{0} \cdot B(0) + \binom{3}{1} \cdot B(1) + \binom{3}{2} \cdot B(2) + \binom{3}{3} \cdot B(3).
\end{aligned}
$$

To compare this with the computation via the corresponding Stirling subset numbers $s_2(4,k)$ $[1 \leq k \leq 4]$, we discuss the following scenario: We have 4 letters to get delivered by 1, 2, 3 or 4 postwomen. (It does not matter who carries the stuff—just which letters are together.) Once we put all letters in just 1 bag, we have only 1 possibility. Taking 2 bags, we have 7 possibilities. Using 3 bags, we can divide the 4 letters in 6 different ways. For employing all 4 bags, only 1 realization remains: all letters are carried in isolation bags. Hence, we obtain the result already known: $B(4) = \sum_{k:=0}^{4} s_2(4,k) = 0+1+7+6+1 = 15$.

The Bell numbers grow very fast:

$$
\begin{aligned}
B(14) &= 190{,}899{,}322, \\
B(15) &= 1{,}382{,}958{,}545,
\end{aligned}
$$

exponentially:

$$2^{(n-1)} \leq B(n) \leq n!;$$

the following holds:

$$4 < n < 2^n < B(n) < n!.$$

Figure 5.9 now illustrates the recursive procedure as a stepwise addition.

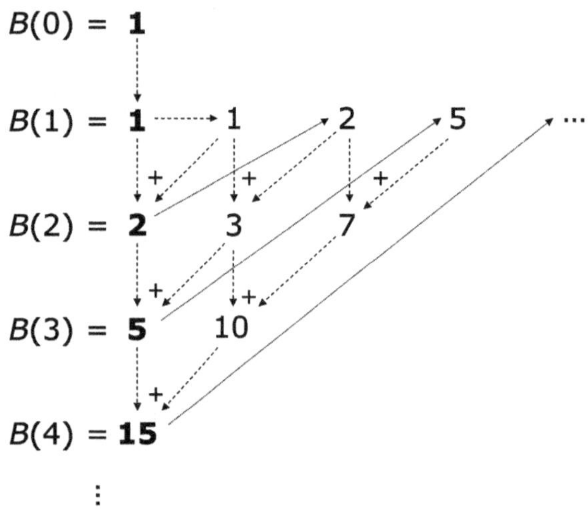


**Figure 5.9:** Bell numbers.

# 6 Probability theory

This chapter thematically concludes the present book. The first section presents the *general probability*, in the second one, we treat the *conditional probability*. We start with some basics (maybe well known), garnished with examples. Afterwards, two versions of a formula by Thomas Bayes come into the picture, each one with an illustrative application: the first one on Roulette and the second one on the difference between theory and practice. ⌣̈ Have fun!

## 6.1 General probability

We introduce the notions needed and explain them along usual examples.

In order to start, we define the *event space*:

$$\Omega := \text{set of all possible occurrences of events}^1 \text{ (see below).}$$

The # elements in this set of possibilities is the *size of the event space*:

$$|\Omega|.$$

A successful *event* is a subset of all possibilities:

$$E \quad [\subseteq \Omega].$$

The *cardinality of an event* therefore is

$$|E|_{[\leq|\Omega|]}.$$

An *elementary event* is an event singleton[2] ($\exists!$ element: $|E| = 1$).
Providing a fair outcome of a so-called "experiment", we yield
the *general probability*:

$$p_\Omega(E) = \frac{|E|}{|\Omega|}.$$

This automatically implies that the probability $p$ ranges between 0 and 1:

$$\begin{aligned} 0\,\% = 0 = p(\{\}) \leq p(E) \leq p(\Omega) = 1 = 100\,\% \\ =_{[\Omega \ni e_i]} p\left(\bigcup_{i:=1}^{|\Omega|} \{e_i\}\right) =_{\substack{\texttt{partition} \\ e_i\ \texttt{disjoint}}} \sum_{i:=1}^{|\Omega|} p(e_i). \end{aligned}$$

1 Of everything which might happen.

2 ⟶ Chapter *Set theory*.

https://doi.org/10.1515/9783111206899-007

Considering $n$ event sets, which might have some elements in common, we realize *Boole's inequality*:

$$p\left(\bigcup_{i:=1}^{n} E_i\right) \leq \sum_{i:=1}^{n} p(E_i).$$

In the general case of the union of potentially nondisjoint event sets, we turn to the counting technique *in/exclusion*; considering just 2 sets, we have the following:

$$p(A \cup B) = p(A) + p(B) - p(A \cap B).$$

In the special case of disjoints sets, the intersection is empty;[3] a partition is given (see above), and we can argue in an intersection-free manner:

$$p(A \cup B) = p(A) + p(B).$$

Once our entire $\Omega$ is divisible in 2 completely independent subareas, we therefore yield

$$\begin{aligned} &p(\Omega) =_{\texttt{partition}} p(A) + p(B) = 1 \quad | - p(B) \quad \Longleftrightarrow \\ &p(A) = 1 - p(B). \end{aligned}$$

Example: Fair coin toss ($H := Head$, $T := Tails$)

$$\Omega := \{H, T\};$$

$$p(H) = 1 - p(T) =_{\texttt{fair}} 1 - p(H) \quad \overset{+p(H)}{\underset{:2}{\Longleftrightarrow}} \quad p(H) = \frac{1}{2} = p(T).$$

The following four examples from "day-to-day" life illustrate larger partitions:

1. 1 dice
   - $\Omega := \{1, 2, 3, 4, 5, 6\}$
   - $|\Omega| = 6$
   - $E :=$ desired number of dots
   - $|E| = 1$
   - $p_\Omega(E) = p(e_i) = |E|/|\Omega| = 1/6, \forall_{[1\leq]} i_{[\leq|\Omega|]}$
   - $p(\Omega) = p(\bigcup_{i:=1}^{|\Omega|}\{e_i\}) = \sum_{i:=1}^{|\Omega|} p(e_i) = 6 \cdot 1/6 = 1$
2. 2 different-colored dice
   - $\Omega := \{(1,1), (1,2), (1,3), \ldots, (1,6), (2,1), \ldots, (2,6), \ldots, (6,1), \ldots, (6,6)\}$
   - $|\Omega| = 6^2 = 36$
   - $E :=$ specific dice toss,[4] $|E| = 1$

3 The reason why we do not have something to subtract.

4 $(j,k) \neq_{[j\neq k]} (k,j)$.

   - $p_\Omega(E) = p(e_i) = |E|/|\Omega| = 1/36, \forall_{[1\le]} i_{[\le|\Omega|]}$
   - $p(\Omega) = p(\bigcup_{i:=1}^{|\Omega|}\{e_i\}) = \sum_{i:=1}^{|\Omega|} p(e_i) = 36 \cdot 1/36 = 1$
3. Pair of (identical) dice
   - $\Omega := \{(j,k) \mid 1 \le j,k \le 6\}$
   - $|\Omega| = 6^2 = 36$—everything as it is
   - $E := \{(j,k) \mid 1 \le j = k \le 6\} = \{(j,j) \mid 1 \le j \le 6\}$
   - $|E| = 6$
   - $p_\Omega(E) = p(\bigcup_{i:=1}^{|E|}\{e_i\}) = \sum_{i:=1}^{|E|} p(e_i) = 6 \cdot 1/36 = |E|/|\Omega| = 1/6$
4. Gambling machine[5] (see Figure 6.1)

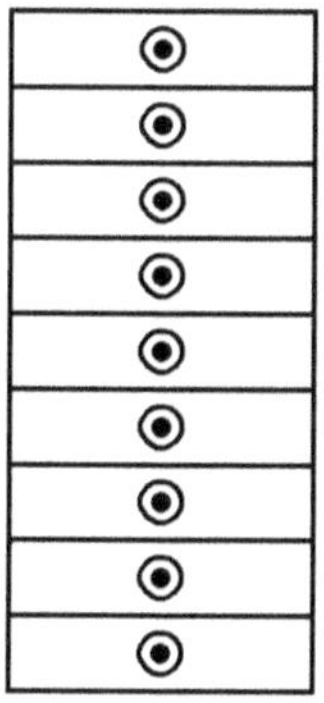

**Figure 6.1:** Gambling machine.

   - $\Omega := \{(b_{n-1}, b_{n-2}, \ldots, b_0) \mid b_i \in \mathcal{B}\ [:= \{0,1\}],\ n-1 \ge i \ge 0\} \quad [= \mathcal{B}^n]$
   - $|\Omega| = |\mathcal{B}|^{|\{n-1,n-2,\ldots,0\}|} = 2^n \quad [= |\mathcal{B}^n|]$
   - $E := \{(1,1,\ldots,1)\}$
   - $|E| = 1$
   - $p_\Omega(E) = |E|/|\Omega| = 1/2^n = 2^0/2^n = 2^{-n}$.

   Concretely: Given a fair *Boole*an button; in order to win a so-called “gambling series” by obtaining 9 (=: $n$) × *true*, we end up by a probability ($\approx_{[<]}$ 0,2 %) close to 2 ‰—a value which is expectable in the environment of locations for such gambling machines. ☺

So far, we have mainly added the probabilities; now, we will multiply them, particularly during the occurrence of “independent” events at the same time.

$n$ events are *independent*, once the following holds:

$$p\left(\bigcap_{i:=1}^{k} E_{j_i}\right) = \prod_{i:=1}^{k} p(E_{j_i}), \quad \forall_{[2\le]} k \le n \ge j_k > j_{k-1} > \cdots > j_2 > j_1 \ge 1.$$

---

5 Simplified view in the sense of a bit-vector, where subsequently each bit must show *true*; pressing $n$ times the gambling button must be successful each time up to the final one.

Illustration:
- $n := 2$

$$p(E_1 \cap E_2) = p(E_1) \cdot p(E_2)$$

- $n := 3$

$$\begin{aligned} p(E_1 \cap E_2) &= p(E_1) \cdot p(E_2) \\ p(E_1 \cap E_3) &= p(E_1) \cdot p(E_3) \\ p(E_2 \cap E_3) &= p(E_2) \cdot p(E_3) \\ p(E_1 \cap E_2 \cap E_3) &= p(E_1) \cdot p(E_2) \cdot p(E_3). \end{aligned}$$

Example: 2 different dice (as previous)
- $\Omega := \{1, 2, 3, 4, 5, 6\}$
- $|\Omega| = 6$ (per dice)
- $E_i :=$ desired dots on the dice$_i$, $1 \leq i \leq 2 =: n =: k$
- $E := E_1 \cap E_2$
- $p_\Omega(E) = p(\{e_1\} \cap \{e_2\}) = p(e_1) \cdot p(e_2) = (1/6)^2 = 1/6^2 = 1/36.$

## 6.2 Conditional probability

This section treats the probability of an event $A$ under the *condition* of a given event $B$; this is calculated as follows—reported: "$p$ of $A$ given $B$":

$$p(A|B) = \frac{p(A \cap B)}{p(B)_{[>0]}} \iff p(A \cap B) = p(A|B) \cdot p(B).$$

In the case of $A$ and $B$ being independent, we yield the unsurprising result:

$$p(A|B) = \frac{p(A) \cdot p(B)_{[>0]}}{p(B)_{[>0]}} = p(A).$$

(The occurrence of $A$ does not depend on the condition of $B$.)

A simple change of the names yields corresponding statements:

$$p(B|A) = \frac{p(B \cap A)}{p(A)_{[>0]}} \iff p(B \cap A) = p(B|A) \cdot p(A).$$

In the case of $A$ and $B$ being independent, we obtain the corresponding formulae:

$$p(B|A) = \frac{p(B) \cdot p(A)_{[>0]}}{p(A)_{[>0]}} = p(B).$$

We now reach the formula of Thomas Bayes related to the *conditional probability*:

$$p(A|B) = \frac{p(A \cap B)}{p(B)_{[>0]}} = \frac{p(B \cap A)}{p(B)_{[>0]}} = \frac{p(B|A) \cdot p(A)}{p(B)}.$$

In the special case $A \subseteq B$ it reads easier:

$$\begin{aligned} p(A|B) &= \frac{p(A \cap B)}{p(B)_{[>0]}} = \frac{p(A)}{p(B)} \quad \Longleftrightarrow \\ p(A \cap B) &= \frac{p(A)}{p(B)_{[>0]}} \cdot p(B)_{[>0]} = p(A); \\ p(B|A) &= \frac{p(B \cap A)}{p(A)_{[>0]}} = \frac{p(A)}{p(A)_{[>0]}} = 1 \quad \Longleftrightarrow \\ p(B \cap A) &= 1 \cdot p(A)_{[>0]} = p(A). \end{aligned}$$

$p(B|A_{[\subseteq B]}) = 100\,\%$, because the more improbable event set $A$ is given and already fulfilled, and the superset $B$ as disjunction of events (OR-connected) is fulfilled anyhow.

This specific *conditional subset probability* could also get inferred in the following way:

$$p(A|B) = p(B|A) \cdot \frac{p(A)_{[>0]}}{p(B)_{[>0]}} = 1 \cdot \frac{p(A)_{[>0]}}{p(B)_{[>0]}} = \frac{p(A)}{p(B)},$$

as already stated above.

We now reach the *total probabilities*:

Precondition: Conditional events $B_1, B_2, \ldots, B_n$ form a partition of $\Omega$:

$$\begin{aligned} &\bigcup_{i:=1}^{n} B_i = \Omega, \\ &B_j \cap B_k = \{\}, \quad \forall_{[1\leq]} j < k_{[\leq n]}; \\ &\sum_{i:=1}^{n} p(B_i) = p(\Omega) = 1. \end{aligned}$$

Due to the fact that $A \subseteq \Omega =_{\texttt{partition}} \bigcup_{i:=1}^{n} B_i$ is given, we obtain $p(A)$ as sum of all $n$ single probabilities of $A$ among the disjoint $B_i$ in the corresponding step:

$$p(A) = \sum_{i:=1}^{n} p(A \cap B_i) = \sum_{i:=1}^{n} (p(A|B_i) \cdot p(B_i)).$$

We now present the general form for the *conditional total probabilities*:

Precondition: $A \subseteq \Omega \supseteq B_s \in \{B_1, B_2, \ldots, B_n\}$; altogether, we obtain the following relationship:

$$p(B_s|A) = \frac{p(B_s \cap A)}{p(A)} = \frac{p(A \cap B_s)}{p(A)} = \frac{p(A|B_s) \cdot p(B_s)}{p(A)} = \frac{p(A|B_s) \cdot p(B_s)}{\sum_{i:=1}^{n} (p(A|B_i) \cdot p(B_i))}.$$

In the special case $n := 1_{[=s]}$, we yield the following 100 % conditional total probability:

$$p(B_s|A) = \frac{p(A|B_s) \cdot p(B_s)}{\sum_{i:=1}^{1}(p(A|B_i) \cdot p(B_i))} = \frac{p(A|B_s) \cdot p(B_s)}{p(A|B_s) \cdot p(B_s)} = 1,$$

which we already get by the special conditional probability case for $A \subseteq B_{(s)}$:

$$p(B_s|A) = p(\Omega|A_{[\subseteq\Omega]}) = 1.$$

As already announced, two examples round off the spectrum of the present book—an obvious one and another one which seems to be nonobvious[6]:

- Example 1: Roulette
  Neglecting the *Zero* we select a color.
  - First run: *Rouge*; second run: Game.
  - By which probability can we expect *Noir*?
  - $A := \{(\underline{R}, N)\}, B := \{(\underline{R}, N), (\underline{R}, R)\} =_{\text{here}} \Omega$.
  - $p(A_{[\subseteq B]}|B) = \frac{p(A)}{p(B)} = \frac{50\,\%}{100\,\%} = \frac{1}{2}$.
  - By which probability can we expect $R$ again?
  - $p(\{(\underline{R}, R)\}_{[\subseteq B]}|\Omega) = \frac{50\,\%}{100\,\%} = \frac{1}{2}$.
  - We realize that both colors are equally probable; we cannot predict the event which follows—which should not be surprising; the outcome of the first run has no effect on the outcome of the second one.
- Example 2: "Behind the door (Monty Hall problem)"
  In the context of an interactive game, some prize is hidden behind 1 of 3 doors. The candidate should now guess the door where the prize is located.
  - Set of doors $D := \{A, B, C\}$
  - Event $H_X$: the prize is situated be*h*ind the door $X$ ($\in D$)
  - Probability $p(H_X) = \frac{1}{3}, \forall X_{[\in D]}$ (equally distributed—"fair")
  - Event $O_X$: door $X$ open
  - The lady starts—and guesses (correctly):[7] $H_A$ (the presented scenario here).
  - The main character of this game is that it comprises two rounds; hence, the host opens 1 of the 2 doors, which the lady has not announced[8]—offering her to reconsider her decision. (The host never opens the door guessed at the first run [independent of being correct or not].) When the prize is really located behind $A$, the host actually has two choices (in order to arbitrarily open) $B$ or $C$; once the prize is placed behind $B$ or $C$, he opens the other one ($C$ or $B$) of course, but not $A$. Here, we let him open $B$.

6 At first glance—revealing the difference between theory and practice ☺.

7 Which she cannot know at that time.

8 A door behind which the prize is not hidden.

- What appears to be more favorable for the lady—to stay with her original choice $A$ or to change to $C$? Or does it even matter?
- She thinks about the reason why door $B$ has been opened (and not $C$):[9]

$$
\begin{aligned}
p(O_B|H_C) &= 1,\\
p(O_B|H_B) &= 0,\\
p(O_B|H_A) &= p(O_C|H_A) = \frac{1}{2}, \quad p(O_A|H_A) = 0;\\
\sum_{X\in D} p(O_X|H_A) &= 0 + \frac{1}{2}\cdot 2 = 1, \quad \text{ok.}
\end{aligned}
$$

- She checks the preconditions of the partitioning of the event space:

$$
\begin{aligned}
\bigcup_{X\in D} H_X &= \Omega,\\
H_A \cap H_B &= H_A \cap H_C = H_B \cap H_C = \{\};\\
\sum_{X\in D} p(H_X) &= \frac{1}{3}\cdot 3 = 1 = p(\Omega), \quad \text{ok.}
\end{aligned}
$$

(After all, the prize is located behind 1 specific door, but not behind several ones; once we could open all doors, the prize would be "safe"—by 100 %.)
- The candidate calculates the conditional total probabilities—in order to prepare her second/final guess: $A$ (to stay with) or $C$ (changing):

$$
\begin{aligned}
p(H_A|O_B) &= \frac{p(O_B|H_A)\cdot p(H_A)}{\sum_{X\in D}(p(O_B|H_X)\cdot p(H_X))} = \frac{\frac{1}{2}\cdot\frac{1}{3}}{(\frac{1}{2}+0+1)\cdot\frac{1}{3}} = \frac{1/2}{3/2} = \frac{1}{2}\cdot\frac{2}{3} = \frac{1}{3};\\
p(H_C|O_B) &= \frac{p(O_B|H_C)\cdot p(H_C)}{\sum_{X\in D}(p(O_B|H_X)\cdot p(H_X))} = \frac{1\cdot\frac{1}{3}}{\frac{3}{2}\cdot\frac{1}{3}} = \frac{2}{3} = 2\cdot p(H_A|O_B).
\end{aligned}
$$

- The lady is enthusiastic and goes "all in"; she lets herself get carried away—toward the change of her attitude.[10] Showdown: the host opens, as finally announced, $C$; the prize, however, is not behind that door: $\neg H_C$, $\neg H_B$; $H_A$.
- Are all cases of partitions considered (after the opening of door $B$)? Voilà:

$$
\sum_{X\in D} p(H_X|O_B) = p(H_A|O_B) + p(H_B|O_B) + p(H_C|O_B) = \frac{1}{3} + 0 + \frac{2}{3} = 1.
$$

- The lady was tough and gave her all, the host as well;
*Discrete Mathematics—Combinatorics, Counting, Proofs, Recurrences, Solutions*—still leaves a gap open for the refreshing difference between theory and practice. ◡̈

---

9 $A$ is not possible due to the rules of the game.

10 To stay with the original one would be more advantageous; see the text commented in footnote 7.

# A Appendix: Exercises

## A.1 Foundations

### Task 1-1

- Functions
  1. Please indicate regarding the following mappings whether they represent functions at all—and in case whether they are even bijective, or at least injective or surjective ("main thing: objective" ◡̈) and justify your answers:

     $$D := \{0, 1, 2\}, \quad C := \{\alpha, \beta, \gamma, \delta\}; \quad f_i : D \to C, \quad 1 \le i \le 3.$$

     (a) $f_1(0) := \alpha, f_1(1) := \beta, f_1(2) := \alpha$
     (b) $f_2(0) := \alpha, f_2(1) := \beta, f_2(2) := \gamma$
     (c) $f_3(0) := \alpha, f_3(1) := \beta, f_3(2) := \gamma, f_3(0) := \delta$
  2. Can a function be bijective in the following case: $|D| \neq |C|$—why (not ◡̈)?
  3. Calculate these values:
     a) $\lceil 0/1 \rceil$
     b) $\lfloor 0/1 \rfloor$
     c) $\lceil 1/1 \rceil$
     d) $\lfloor 1/1 \rfloor$
     e) $\lfloor 0/1 \rceil$
     f) $\lfloor 1/1 \rceil$
     g) $\lceil 1/2 \rceil$
     h) $\lfloor 1/2 \rfloor$
- Relations
  1. $X := \{0, 1, 2\}, Y := \{1, 2, 3\}$. Determine the following (each time $\subset X \times Y$):
     a) $<(X, Y)$
     b) $>(X, Y)$
  2. $\mathcal{B} := \{0, 1\}$. Which formula value does $|\mathcal{B}^n|$ produce?

### Solution 1-1

- Functions
  1. The first mapping is a usual one without any further feature; the second one is still not surjective (and therefore not bijective), but injective. The third construction is not a function at all—an objective ◡̈ situation:
     (a) The function $f_1$ is not a specific one in a certain sense:
        i. $f_1(0) = f_1(2) \Longrightarrow \neg$ injective $\Longrightarrow \neg$ bijective
        ii. $|C| > |D| \Longrightarrow \neg$ surjective ($\Longrightarrow \neg$ bi . . . ◡̈)

https://doi.org/10.1515/9783111206899-008

(b) $f_2$ is injective, but not surjective:
  i. All function values are different $\Longrightarrow f_2$ injective
  ii. $\neg$ surjective (see above) $\Longrightarrow \neg$ bijective

(c) The intended mapping is nonsense: $\alpha =: f_3(0) := \delta$.

2. A function with $|D| \neq |C|$ can never be bijective:
   (a) bijective $\Longleftrightarrow$ injective $\wedge$ surjective
   (b) injective $\Longrightarrow |D| \leq |C|$
   (c) surjective $\Longrightarrow |D| \geq |C|$
   (d) bijective $\Longrightarrow |D| \leq |C| \leq |D| \Longleftrightarrow |D| = |C|$
   (e) contrapositive: $|D| \neq |C| \Longrightarrow \neg$ bijective
3. In the first 6 fractions nothing has to get rounded, just the final 2:
   a) 0
   b) 0
   c) 1
   d) 1
   e) 0
   f) 1
   g) 1
   h) 0

- Relations
  1. The set of relations are proper subsets of the Cartesian Product ($CP$):
     (a) $\{(0,1),(0,2),(0,3),(1,2),(1,3),(2,3)\}$
     (b) $\{(2,1)\}$ with (casually noted) $|(b)| = 1 < 6 = |(a)| < |CP| = 3^2 = 9$
  2. $|\mathcal{B}^n| = |\mathcal{B}|^n = 2^n$

### Task 1-2

Given are the bijections $f : A \to B, g : B \to C, h := g \circ f; |A| =: a, |B| =: b, |C| =: c_{[>0]}$.
How many bijective compositions are possible for $h$?
Which typical mistakes lurk—and why is it nevertheless simple not to fall for it? ◡̈

### Solution 1-2

$a!$ ($= b!$ [$= c!$]); Proof: starting on page 67.
Illustration: $a := 3$ ($= c$); $A := \{\alpha, \beta, \gamma\}$, $C := \{x, y, z\}$.
# bijections = 3! = 6:

$$h_1\colon\quad h_1(\alpha) := x,\quad h_1(\beta) := y,\quad h_1(\gamma) := z;$$
$$h_2\colon\quad h_2(\alpha) := x,\quad h_2(\beta) := z,\quad h_2(\gamma) := y;$$
$$h_3\colon\quad h_3(\alpha) := y,\quad h_3(\beta) := x,\quad h_3(\gamma) := z;$$
$$h_4\colon\quad h_4(\alpha) := y,\quad h_4(\beta) := z,\quad h_4(\gamma) := x;$$

$$h_5: \quad h_5(\alpha) := z, \quad h_5(\beta) := x, \quad h_5(\gamma) := y;$$
$$h_6: \quad h_6(\alpha) := z, \quad h_6(\beta) := y, \quad h_6(\gamma) := x.$$

The following misunderstandings are notorious set-breaking points:

- The bijection per se: $\neq |C|^2$ ($c \neq 1$, of course).
  For $2 \leq c \leq 3$, the correct solution is smaller than the one presented here, because by the square idea all elements are still considered to be accessible; it is greater in the case $c \geq 4$, because now the full permutation explosion applies (producing an exponentially large number [where the indicated number 4 plays the role of the $n_0$ in an induction proof]).
- To consider the composition too complicated:
  One could first think about to determine the combinatorics of the # $f$-bijections (=: $k$) and building on that the # $g$-bijections (=: $l$), in order to get the total # $h$-bijections (=: $m$), for instance in the following manner:

$$m := l := k! := c!! >_{[c>2]} c! \quad —$$

  maybe tempting, but spooky.
  The easiest way to grasp it is to immediately take into account—as performed here—the effect of a composition on the original function $f$ (neglecting the intermediate function [$g$] ◡̈) on the # output possibilities of the function $h$; thereby, the permutation drama could not get further dramatized. ($a = b = c$; otherwise, a bijection is impossible; the combinatorics of a bijection is just based on how often the elements in the [co]domain can get numbered consecutively.)

**Task 1-3**

Given a *function* $f$, which starts with a $d$-elementary *domain* to produce values from a smaller $c$-elementary *codomain*: Is $f$ necessarily surjective?

Illustrate a possible scenario to support the correct ◡̈ answer!

**Solution 1-3**

$f$ <u>not</u> necessarily surjective: $d := 3 > 2 =: c$; $D := \{Q_1, Q_2, Q_3\}$, $C := \{N, Y\}$:

$$f(Q_1) := N =: f(Q_2) =: f(Q_3) \neq_{:(} Y_{[\in C]} \notin \{f(Q_i) \mid 1 \leq i \leq 3\}$$

**Task 1-4**

Given a *function* $f$, which starts with a $d$-elementary *domain* to produce values from a larger $c$-elementary *co-domain*: Is $f$ necessarily injective?

Illustrate a possible scenario to support the correct ◡̈ answer.

**Solution 1-4**

$f$ not necessarily injective: $d := 2 < 3 =: c$:
$D :=$ {decision-1, decision-2}, $C :=$ {yes, no, indifferent}
$f$(decision-1) := no =: $f$(decision-2)
[($D \ni$) decision-1 $\neq$ decision-2 ($\in D$)]

**Task 1-5**

Let a function $f$ work on the *d*omain $D$, which is larger than the *c*odomain $C$:
How does the pigeonhole principle show that $f$ cannot be injective?

**Solution 1-5**

There exists at least 1 codomain value, which gets selected by $\lceil \frac{|D|}{|C|} \rceil \geq 2$ domain entries.

**Task 1-6**

51 students write their exam in 4 rooms: What does the pigeonhole principle tell us?
Which test calculation would easily exclude the opposite?

**Solution 1-6**

$\exists$ at least 1 room with at least $\lceil 51/4 \rceil = 13$ students.

The opposite would be that 12 places are sufficient: $12 \cdot 4 = 48 < 51$, which shows that having only 12 seats in each room are not sufficient; 3 stud. would not have a seat.[1] :(

**Task 1-7**

Which condition must exist for the following statement: $|\bigcup_{i:=1}^{n} S_i| = \sum_{i:=1}^{n} |S_i|$?

**Solution 1-7**

All $S_i$-pairs are mutually disjoint.

1 This might not be a problem ;) Students may enter the exam room, have a look on the questions, and immediately leave the setting.

## A.2 Set theory

### Task 2-1

- Notions/Cardinality of finite sets
  1. Construct two finite sets $A$ and $B$ with at least 1 common element (non-empty intersection), s. t. the following holds: $|A| = |B| + 2$.
     (a) Visualize by a drawing the cardinality of the set union
     $|A \cup B| := |A| + |B| - |A \cap B|$.
     Argue why we must subtract the intersection elements $\underline{1}\times$.
     (b) Calculate $|A n B|$ via the corresponding formula $(|A| - |A \cap B|)$ and compare your answer with the procedure to immediately form the set difference in order to count only these (remaining) elements.
     (c) Calculate $|A \oplus B|$ via the known (◡̈) formula and compare this answer with the procedure to form the symmetric difference directly and to count these elements only.
  2. Given is the set $S := \{1, 2, 3, 4, \alpha, \beta, \gamma, \mathit{false}, \mathit{true}\}$; we now produce the *partition* $P := \{A_1, A_2, A_3\}$ with $A_1 := \{1, 2, 3, 4\}$, $A_2 := \{\alpha, \beta, \gamma\}$, $A_3 := \{\mathit{false}, \mathit{true}\}$.
     Determine:
     a) $|P|$
     b) $|S|$
- Laws
  Let our small *universe* be the world of decimal numbers: $U := \{0, 1, 2, \dots, 9\}$.
  1. $E := \{0, 2, 4, 6, 8\}$.
     Construct:
     a) $E^c$
     b) $(E^c)^c$
  2. In this task, please produce all expressions of the two variants of the dual *DeMorgan* principle and enlighten the relationship in the corresponding law: $n := 3$; $S_1 := \{1, 2, 3, 4, 5\}$, $S_2 := \{2, 4, 6, 8\}$, $S_3 := \{3, 6, 9\}$.
- Un/countability of infinite sets
  How are we able to construct a set having a higher cardinality compared to a given set $S$ even when $S$ is infinite already?
  (This would imply that an infinite number of infinity levels exist.)

### Solution 2-1

- Notions/Cardinality of finite sets
  1. It is interesting for 2 sets, which are neither sub- nor superset of each other. Let us take $A := E$, the set defined beforehand, and $B := \{1, 6, 9\}$.
     (a) You create, of course, your personal illustration. ◡̈

For the sake of convenience, let us use $A \cap B =: T$. We see the following:

$$|A \cup B| = |A \cup (B \setminus (A \cap B))| =_{\texttt{partition}} |A| + (|B \setminus (A \cap B)|)$$
$$=_{B \supset T} |A| + (|B| - |A \cap B|) =_{\texttt{rule of associativity}} |A \cup B| \quad [= 7].$$

By the $|\cup|$ calculation, the intersection $T$ is (virtually) considered $2\times$ ($T$ is part of both sets given)—the reason why we must subtract $1 \times |\cap|$.

(b) Here, it is intended that you first evaluate the formula concretely and then to compare the result with the direct counting of the elements exclusively in the set of difference, which should result in the same outcome. [In the example above, you should end up with 4.]

(c) Similar to "(b)," now related to the symmetric difference [= 6].

2. Realizing a 3-parts partition $P$, the base set $S$ gets partitioned in three parts; $S$ (with its nine elements), of course, keeps its base cardinality:
   (a) $|P| = 3$
   (b) $|S| = \sum_{i:=1}^{|P|} |A_i| = 4 + 3 + 2 = 9$

- Laws
  This item treats the elements, which do not belong to a given set.
  1. Given $E$ := set of the (five) even decimal numbers.
     (a) Hence, the other 5 odd decimal numbers (=: $O$) are missing.
     (b) Once we complement this set $O$ ("back"), we end up by the original set ($E$) again; therefore, there is no need to perform this double complement computationally, which is the idea to mention here. ☺
  2. We show it for the complement[2] of the $\cap$- as well as of the $\cup$-set:
     - Complement of the total intersection = union of all single complements:
       i. $(\bigcap_{i:=1}^{3} S_i)^c = \{\}^c = U$
       ii. $\bigcup_{i:=1}^{3} S_i^c = \{0,6,7,8,9\} \cup \{0,1,3,5,7,9\} \cup \{0,1,2,4,5,7,8\} = U$
     - Complement of the total union = intersection of all single complements:
       i. $(\bigcup_{i:=1}^{3} S_i)^c = \{1,2,3,4,5,6,8,9\}^c = \{0,7\}$
       ii. $\bigcap_{i:=1}^{3} S_i^c = \{0,6,7,8,9\} \cap \{0,1,3,5,7,9\} \cap \{0,1,2,4,5,7,8\} = \{0,7\}$
- Un/countability of infinite sets
  $|\mathcal{P}(\omega_i)| >_{[i \geq 1,\ \texttt{(generalized) continuum hypothesis}]} \omega_i$

## Task 2-2

Please complete: $|M_1| = |M_2| \iff \cdots$

## Solution 2-2

$\exists$ bijection: $M_1 \to M_2$

2 Without "i" (in the middle) ☺.

### Task 2-3

Construct a 9-elementary set $P$, form a partition with 3 sets $A$, $B$, and $C$ of different sizes and thereby illustrate a special case for the calculation of the cardinality of the union of several sets! Which cumbersome principle is not necessary here?

### Solution 2-3

$P := \{1, 2, 3, \ldots, 9\}; A := \{1, 2, 4, 8\}, B := \{3, 6, 9\}, C := \{5, 7\}.$
$|P| := |A \cup B \cup C| = |A| + |B| + |C|.$

Due to the feature of a partition having all 3 single sets being disjunct the in/exclusion principle is not needed.

### Task 2-4

1. The cardinality of the *power-set* $|P(S)| > |S|$.
   How can we illustrate this—in a finite as well as in an infinite setting?
2. In which scenario can a set $S$ have a proper subset $T$ with $|S| = |T|$, in which one not? Support your statement(s).

### Solution 2-4

1. In the finite setting, the following holds: $|M| =: k, |P(M)| = 2^k > k$.
   Each element could be written inside a pair of parentheses, additionally to the empty set (">"); a usual induction prove yields the exact dependence on the doubling effect ("=").
   In an infinite setting, the famous continuum hypothesis enters the picture.
2. For infinite sets, this is even their definition. For finite sets, it is impossible: at least 1 element is missing.

## A.3 *Boole*an algebra

### Task 3-1

- Truth tables and logical combinatorics
  1. Show the following equivalences:
     (a) $p \oplus q \Longleftrightarrow (p \vee q) \wedge (p \mid q)$
     (b) *Implication* $\Longleftrightarrow \neg p \vee q \Longleftrightarrow$ *Contrapositive*
     (c) *Converse* $\Longleftrightarrow$ *Inverse*
     (d) *Equivalence* $\Longleftrightarrow$ *Implication* AND *Converse*

2. Given $n$ *Boole*an variables; justify the corresponding standard formula, for
   (a) # different codings (combinatorics)
   (b) # different functions

– Laws
  Please prove, even in both variants if needed:
  1. Absorption
  2. *De Morgan*
  3. Exportation

## Solution 3-1

– Truth tables and logical combinatorics
  1. Equivalences via corresponding logic table:

| $p$ | $q$ | $p \oplus q$ | $p \vee q$ | $p \mid q$ | $(p \vee q) \wedge (p \mid q)$ |
|---|---|---|---|---|---|
| 0 | 0 | **0** | 0 | 1 | **0** |
| 0 | 1 | **1** | 1 | 1 | **1** |
| 1 | 0 | **1** | 1 | 1 | **1** |
| 1 | 1 | **0** | 1 | 0 | **0** |

| $p$ | $q$ | $p \rightarrow q$ | $\neg p$ | $\neg p \vee q$ | $\neg q$ | $\neg q \rightarrow \neg p$ |
|---|---|---|---|---|---|---|
| 0 | 0 | **1** | 1 | **1** | 1 | **1** |
| 0 | 1 | **1** | 1 | **1** | 0 | **1** |
| 1 | 0 | **0** | 0 | **0** | 1 | **0** |
| 1 | 1 | **1** | 0 | **1** | 0 | **1** |

  (a) $p \oplus q \Longleftrightarrow (p \vee q) \wedge (p \mid q)$
  (b) *Implication* $\Longleftrightarrow \neg p \vee q \Longleftrightarrow$ *Contrapositive*
  (c) *Converse* $\Longleftrightarrow$ *Inverse*
      Similar to *Implication* $\Longleftrightarrow$ *Contrapositive*, now just with $p$ and $q$ swapped.
  (d) *Equivalence* $\Longleftrightarrow$ *Implication* AND *Converse*

| $p$ | $q$ | $p \longleftrightarrow q$ | $p \rightarrow q$ | $q \rightarrow p$ | $(p \rightarrow q) \wedge (q \rightarrow p)$ |
|---|---|---|---|---|---|
| 0 | 0 | **1** | 1 | 1 | **1** |
| 0 | 1 | **0** | 1 | 0 | **0** |
| 1 | 0 | **0** | 0 | 1 | **0** |
| 1 | 1 | **1** | 1 | 1 | **1** |

  2. Here come the explanations of the requested formulae for $n$ *Boole*an variables:
     (a) # different codings (combinatorics)
         The cardinality of this specific Cartesian product on the common base set $\mathcal{B}$ is $2^n$, which is actually shown in detail on page 8.

(b) # different functions

Each of the $2^n$ assignments from "(a)" represents a concrete realization, which gets evaluated in a formula[3] to `true` or `false`—which yields $2^{(2^n)}$ different ($\mathcal{B}$-)functions; page 41 presents the proof of this as nice ⌣̈ exercise on induction.

- Laws

1. Absorption

| $p$ | $q$ | $p \vee q$ | $p \wedge (p \vee q)$ | $p \wedge q$ | $p \vee (p \wedge q)$ |
|---|---|---|---|---|---|
| **0** | 0 | 0 | **0** | 0 | **0** |
| **0** | 1 | 1 | **0** | 0 | **0** |
| **1** | 0 | 1 | **1** | 0 | **1** |
| **1** | 1 | 1 | **1** | 1 | **1** |

2. *De Morgan*

| $p$ | $q$ | $p \wedge q$ | $\neg(p \wedge q)$ | $\neg p$ | $\neg q$ | $\neg p \vee \neg q$ |
|---|---|---|---|---|---|---|
| 0 | 0 | 0 | **1** | 1 | 1 | **1** |
| 0 | 1 | 0 | **1** | 1 | 0 | **1** |
| 1 | 0 | 0 | **1** | 0 | 1 | **1** |
| 1 | 1 | 1 | **0** | 0 | 0 | **0** |

| $p$ | $q$ | $p \vee q$ | $\neg(p \vee q)$ | $\neg p$ | $\neg q$ | $\neg p \wedge \neg q$ |
|---|---|---|---|---|---|---|
| 0 | 0 | 0 | **1** | 1 | 1 | **1** |
| 0 | 1 | 1 | **0** | 1 | 0 | **0** |
| 1 | 0 | 1 | **0** | 0 | 1 | **0** |
| 1 | 1 | 1 | **0** | 0 | 0 | **0** |

3. Exportation

| $p$ | $q$ | $r$ | $q \rightarrow r$ | $p \rightarrow (q \rightarrow r)$ | $p \wedge q$ | $(p \wedge q) \rightarrow r$ |
|---|---|---|---|---|---|---|
| 0 | 0 | 0 | 1 | **1** | 0 | **1** |
| 0 | 0 | 1 | 1 | **1** | 0 | **1** |
| 0 | 1 | 0 | 0 | **1** | 0 | **1** |
| 0 | 1 | 1 | 1 | **1** | 0 | **1** |
| 1 | 0 | 0 | 1 | **1** | 0 | **1** |
| 1 | 0 | 1 | 1 | **1** | 0 | **1** |
| 1 | 1 | 0 | 0 | **0** | 1 | **0** |
| 1 | 1 | 1 | 1 | **1** | 1 | **1** |

## Task 3-2

Show, without truth table, by logical transformations the basis of the indirect proof:

$$p \longrightarrow q \Longleftrightarrow \neg q \longrightarrow \neg p.$$

---

**3** For example, in a CNF (conjunctive *normal form*), which is either *satisfied* or not ($\longrightarrow$ SAT-theory).

**Solution 3-2**

$$p \longrightarrow q$$
$$\Longleftrightarrow$$
$$\neg p \vee q$$
$$\Longleftrightarrow$$
$$q \vee \neg p$$
$$\Longleftrightarrow$$
$$\neg\neg q \vee \neg\neg(\neg p)$$
$$\Longleftrightarrow$$
$$\neg(\neg q) \vee \neg(\neg\neg p)$$
$$\Longleftrightarrow$$
$$\neg(\neg q) \vee \neg p$$
$$\Longleftrightarrow$$
$$\neg q \longrightarrow \neg p$$

**Task 3-3**

Please prove: $p \longleftrightarrow q \Longleftrightarrow (p \wedge q) \oplus (\neg p \wedge \neg q)$

**Solution 3-3**

| $p$ | $q$ | $p \longleftrightarrow q$ | $\neg p$ | $\neg q$ | $(p \wedge q) =: l$ | $(\neg p \wedge \neg q) =: r$ | $l \oplus r$ |
|---|---|---|---|---|---|---|---|
| 0 | 0 | **1** | 1 | 1 | 0 | 1 | **1** |
| 0 | 1 | **0** | 1 | 0 | 0 | 0 | **0** |
| 1 | 0 | **0** | 0 | 1 | 0 | 0 | **0** |
| 1 | 1 | **1** | 0 | 0 | 1 | 0 | **1** |

**Task 3-4**

You could now take the following variant with the absolute value sign[4] or the notation with the [Stirling-]cycle number[5]. Prove by induction on $n_{[>0]}$:

$$z_n := |s_1(n,1)| = (n-1)!.$$

4 Complex :-(.

5 Comfortable (recommendable) ☺.

**Solution 3-4**

Absolute value: $z_n := |s_1(n,1)|$
(i) Basis: $n_0 := 1$

$$z_{1_{\text{principle}}} := \left|s_1(1,1)\right| = |1| = 1$$
$$z_{1_{\text{formula}}} := (1-1)! = 0! = 1 = z_{1_P}.$$

(ii) Hypothesis: $z_{n-1} := |s_1(n-1,1)| = ((n-1)-1)!$
(iii) Step: $n-1_{[\geq n_0]} \to n_{[>n_0]}$;

$$\begin{aligned}
z_{n_P} &:= \left|s_1(n,1)\right| \\
&= \begin{cases} +s_1(n,1); & \text{even}(n-1) \\ -s_1(n,1); & \text{odd}(n-1) \end{cases} \\
&= \begin{cases} +(s_1(n-1,0)-(n-1)\cdot s_1(n-1,1)); & \text{even}(n-1) \\ -(s_1(n-1,0)-(n-1)\cdot s_1(n-1,1)); & \text{odd}(n-1) \end{cases} \\
&= \begin{cases} +(0-(n-1)\cdot s_1(n-1,1)); & \text{even}(n-1) \\ -(0-(n-1)\cdot s_1(n-1,1)); & \text{odd}(n-1) \end{cases} \\
&= (n-1)\cdot \begin{cases} (-s_1(n-1,1)); & \text{odd}((n-1)-1) \\ (+s_1(n-1,1)); & \text{even}((n-1)-1) \end{cases} \\
&= (n-1)\cdot\left|s_1(n-1,1)\right| = (n-1)\cdot z_{n-1} \stackrel{!}{=} (n-1)\cdot((n-1)-1)! \\
&= (n-1)! = z_{n_F}
\end{aligned}$$

Cycle notation: $z_n := \left[{n \atop 1}\right]$
(i) Basis: $n_0 := 1$

$$z_{1_{\text{principle}}} := \begin{bmatrix} 1 \\ 1 \end{bmatrix} = 1$$
$$z_{1_{\text{formula}}} := (1-1)! = 0! = 1 = z_{1_P}.$$

(ii) Hypothesis: $z_{n-1} := \left[{n-1 \atop 1}\right] = ((n-1)-1)!$
(iii) Step: $n-1_{[\geq n_0]} \to n_{[>n_0]}$

$$\begin{aligned}
z_{n_P} &:= \begin{bmatrix} n \\ 1 \end{bmatrix} = \begin{bmatrix} n-1 \\ 0 \end{bmatrix} + (n-1)\cdot\begin{bmatrix} n-1 \\ 1 \end{bmatrix} = 0 + (n-1)\cdot z_{n-1} \\
&\stackrel{!}{=} (n-1)\cdot((n-1)-1)! = (n-1)! = z_{n_F}.
\end{aligned}$$

### Task 3-5

Prove the following statement:

The set of divisors $T_n$ is even [=: $l$] <u>iff</u> [$r$ :=] $n_{[>0]}$ is not a square number.
Show first (a) $r \Rightarrow l$ and then (b) $\neg r \Rightarrow \neg l$ [$\Longleftrightarrow (l \Rightarrow r)$], whereby all together $l \Longleftrightarrow r$.
Finally, describe the situation (c) for a prime number. ⌣̈

### Solution 3-5

$z := |T_n|, m := \lceil \frac{z}{2} \rceil$;

(a) $n$ nonsquare

$T_n := \{1 =: t_1, t_2, t_3, \ldots, t_{m-1}, t_m (\leq \lfloor \sqrt{n} \rfloor), t_{m+1} (> \lfloor \sqrt{n} \rfloor), t_{m+2}, \ldots, t_z := n\}$ with $t_i < t_{i+1}$ for $0 < i < z$.

$t_k | n$; $\{t_k, \frac{n}{t_k}\} =: S_k \subseteq T_n$ with $|S_k| = 2$ for $1 \leq k \leq m = |\{S_1, \ldots, S_m\}|$.

$t_{m+1} = \frac{n}{t_m}, t_{m+2} = \frac{n}{t_{m-1}}, \ldots, t_z = \frac{n}{t_1} (= n)$.

$t_j = \frac{n}{t_k}$ with $z \geq j > m \geq k = z - j + 1 \geq 1$.

$j + k = 2 \cdot m + 1 = j + (z - j + 1) = z + 1 \Longleftrightarrow z + 1 = 2m + 1 \Longleftrightarrow z = 2m$ <u>even</u> ($\geq 2$).

Illustration: $2m = 2 \cdot \lceil \frac{z}{2} \rceil =_{[z \text{ even}]} 2 \cdot \frac{z}{2} = z$.

Test$_1$: $n_1 := 48$ (even):
$z_1 = |T_{48}| = |\{1, 2, 3, 4, 6 (= \lfloor \sqrt{48} \rfloor), 8 (> \lfloor \sqrt{48} \rfloor), 12, 16, 24, 48\}| = 10 = 2 \cdot 5 = 2 \cdot m_1$;

Test$_2$: $n_2 := 69$ (odd):
$z_2 = |T_{69}| = |\{1, 3 (< \lfloor \sqrt{69} \rfloor), 23 (> \lfloor \sqrt{69} \rfloor), 69\}| = 4 = 2 \cdot 2 = 2 \cdot m_2$ even.

(b) Square number $n = (\sqrt{n})^2 = (t_m)^2$, $t_m = \sqrt{n} \mid n$, $t_{m+1} =_? \frac{n}{t_m} = \frac{n}{\sqrt{n}} = \frac{(\sqrt{n})^2}{\sqrt{n}} = \sqrt{n} =_! t_m$; the (standard-)set ($T_n$) however does not contain copies; therefore, $t_m = \sqrt{n} = \frac{n}{\sqrt{n}}$ does not exist twice.
(And also here, the correct $t_{m+1} > t_m$ for $m > 1 < 4 \leq n$.) $\Longrightarrow z = 2 \cdot m - 1$ odd.

Illustration: $m = \frac{z+1}{2} =_{[z \text{ odd}]} \lceil \frac{z}{2} \rceil$.

Test$_0$: $n_0 := 1$ (odd):
$z_0 = |T_1| = |\{1 (= \sqrt{1})\}| = 1 = 2 \cdot 1 - 1 = 2 \cdot m_0 - 1 = 2 \cdot \lceil \frac{1}{2} \rceil - 1$ odd;

Test$_1$: $n_1 := 36$ (even):
$z_1 = |T_{36}| = |\{1, 2, 3, 4, 6 (= \sqrt{36}), 9, 12, 18, 36\}| = 9 = 2 \cdot 5 - 1 = 2 \cdot m_1 - 1$;

Test$_2$: $n_2 := 9$ (odd):
$z_2 = |T_9| = |\{1, 3 (= \sqrt{9}), 9\}| = 3 = 2 \cdot 2 - 1 = 2 \cdot m_2 - 1$ odd.

(c) $n := p$ prime $\Longrightarrow$ (nonsquare) special case of (a): $|T_p| = |\{1, p\}| = 2 = 2 \cdot 1 = 2 \cdot m$ even. ⌣̈

Test$_1$: $p_1 := 2$ (even):
$z_1 = |T_2| = |\{1 (= \lfloor \sqrt{2} \rfloor), 2 (> \lfloor \sqrt{2} \rfloor)\}| = 2 = 2 \cdot 1 = 2 \cdot m_1$;

Test$_2$: $p_2 := 5$ (odd):
$z_2 = |T_5| = |\{1\ (< \lfloor\sqrt{5}\rfloor), 5\ (> \lfloor\sqrt{5}\rfloor)\}| = 2 = 2 \cdot 1 = 2 \cdot m_2$ even.

## A.4 Proof principles

### Task 4-1

- Induction
  Please prove: $n! > 2^n, \forall n \geq n_0$.
- Direct proof
  Derive the Gaussian formula for the sum of the first $n$ natural numbers.
- Indirect proof
  Show that for two natural numbers the following holds: $a < b \Longleftrightarrow a^2 < b^2$.

### Solution 4-1

- Induction
  Statement: $n! > 2^n, \forall n \geq n_0 := 4$.
  Basis: $n_0 := 4$; $4! = 24 > 16 = 2^4$
  Hypothesis [=: H.]: $(n-1)! > 2^{(n-1)}, n - 1 \geq 4$
  Step: $[4 \leq]\ n - 1 \longrightarrow n[> 4]$

$$n! := (n-1)! \cdot n \overset{!}{>}_{[\mathrm{H.}]} 2^{(n-1)} \cdot n >_{[n > n_0 > 2]} 2^{(n-1)} \cdot 2^1 = 2^{[(n-1)+1]} = 2^n.$$

- Direct proof
  To get this Gaussian formula, you could just visit page 43 and exchange $n - 1$ by $n$, or more explicitly—and probably more historically—as follows:

$$\sum_{i:=1}^{n} i = \left[\sum_{i:=1}^{n} i + \sum_{i:=0}^{n-1} (n-i)\right]/2 = \left[\sum_{i:=1}^{n} (i + [n - (i-1)])\right]/2$$
$$= \left[\sum_{i:=1}^{n} (i + n - i + 1)\right]/2 = \left[\sum_{i:=1}^{n} (n+1)\right]/2 = \frac{n \cdot (n+1)}{2}$$

- Indirect proof
  An $l \longleftrightarrow r$-relationship famously ◡̈ looks as follows: $(l \longrightarrow r) \wedge (r \longrightarrow l)$.
  You should be able to show $l \longrightarrow r$ ($\Longleftrightarrow_{\mathtt{contrapositive}} \neg r \longrightarrow \neg l$) ◡̈; hence, let us focus on the second AND-part, which can easily be derived indirectly via its *contrapositive*:

$$\neg(a < b) \Longleftrightarrow a \geq b \Longrightarrow a =_{[\beta \geq 1]} \beta \cdot b \Longrightarrow a^2 = \beta^2 \cdot b^2$$
$$\Longrightarrow_{[\beta^2 \geq 1]} a^2 \geq b^2 \Longleftrightarrow \neg(a^2 < b^2)\ \hat{=}\ \neg l \longrightarrow \neg r.$$

**Task 4-2**

Laurence E. Sigler, Fibonacci's Liber Abaci—A Translation into Modern English of Leonardo Pisano's Book of Calculation, page 397, Springer, 2003, 978-0-387-40737-1 (original probable from 1202):

> *"On Him Who Went into the Pleasure Garden to Collect Apples."*
> A certain man entered a certain pleasure garden through 7 doors, and he took from there a number of apples; when he wished to leave he had to give the first doorkeeper half of all the apples and one more; to the second doorkeeper he gave half of the remaining apples and one more. He gave to the other 5 doorkeepers similarly, and there was one apple left for him.

$$a_d := \#\ \text{apples to pass}\ d\ \text{doorkeepers (s. t. exactly 1 apple remains).}$$

Find the formula for the general case $a_n$ and prove your statement.

**Solution 4-2**

You may consult my German Informatics book[6] (pp. 25/26) cited at the end. ☺

## A.5 Counting techniques

**Task 5-1**

- Pigeonhole principle (=: *PP*)

$$s := \#\ \text{students} := 60, \quad r := \#\ \text{rooms}.$$

  Suppose the *PP* would tell us: There exists a room with at least 9 students.
  (Caution: We do not employ this counting technique now in its usual manner.)
  How many rooms are offered?
- In/exclusion

$$A \cap B \cap C =: D.$$

  Please answer in both tasks the following question:
  Does there an element ($\in D$) exist, which belongs to each set given: $|D| > 0$?
  1. $|A| := 1, |B| := 2, |C| := 3,$
     $|A \cap B| := |B \cap C| := 1, |A \cap C| := 0,$
     $|A \cup B \cup C| := 4$

6 2nd edition work in progress.

2. $|A| := 2, |B| := 3, |C| := 4,$
   $|A \cap B| := |B \cap C| := 2, |A \cap C| := 1,$
   $|A \cup B \cup C| := 5$

- Permutations
  We have the following letters: $A, B, D, E, F, I, W$; both the "$E$" and the "$I$" exist twice. Hence, we have the vowels (=: $V$) $A, E, E, I, I$ and the consonants (=: $C$) $B, D, F, W$. How many visibly different arrangements might be possible to construct a string consisting of these 9 characters once such a word should be of the following form:

$$V \mid C \mid V \mid C \mid V \mid C \mid V \mid C \mid V?$$

### Solution 5-1

- Pigeonhole principle (=: $PP$): $p := \mathrm{PP} - \# := \lceil \frac{s}{r} \rceil$

$$v = \left\lceil \frac{s}{p} \right\rceil =_{\text{here}} \left\lceil \frac{60}{9} \right\rceil = \left\lceil \frac{9 \cdot 6 + 6}{9} \right\rceil = \left\lceil 6\frac{2}{3} \right\rceil = 7$$

Test:

$$\left\lceil \frac{60}{8} \right\rceil = 8 < p = \left\lceil \frac{60}{7} \right\rceil = \left\lceil \frac{7 \cdot 8 + 4}{7} \right\rceil = \left\lceil 8\frac{4}{7} \right\rceil = 9 < 10 = \left\lceil \frac{60}{6} \right\rceil$$

- In/exclusion
  1. $|A \cap C| := 0 \Longrightarrow |D| = 0 \not> 0 \Longrightarrow$
     No, there is not any element in the common intersection set.
  2. $|A \cup B \cup C| = |A| + |B| + |C| - (|A \cap B| + |A \cap C| + |B \cap C|) + |A \cap B \cap C| \Longleftrightarrow$
     $|D| = 5 - (2 + 3 + 4) + (2 + 1 + 2) = 5 - 9 + 5 = 1 > 0 \Longrightarrow$
     Yes, this time there exists such an element.
- Permutations

$$a = \frac{\frac{9!}{1! \cdot 1! \cdot 1! \cdot 1! \cdot 1! \cdot 2! \cdot 2!}}{\frac{9!}{5! \cdot 4!}} = \frac{5! \cdot 4!}{2! \cdot 2!} = \frac{(2! \cdot 3 \cdot 4 \cdot 5) \cdot (2! \cdot 3 \cdot 4)}{2! \cdot 2!} = (3 \cdot 4)^2 \cdot 5 = \frac{1440}{2} = 720$$

or (more compact):

$$\frac{5!}{1! \cdot 2! \cdot 2!} \cdot 4! = \frac{2! \cdot 3 \cdot 4 \cdot 5}{2! \cdot 2!} \cdot 2! \cdot 3 \cdot 4 = 12^2 \cdot 5 = \frac{1440}{2} = 720.$$

### Task 5-2

During a soccer training we have $p_{[>1]}$ player and like to form 2 teams with about the same number of players at each side.

The question in the following 3 cases[7] [a), b), c)] always is:

How many different *formations* [=: $f(p) = \cdots$] are possible?

a) $p := 5$

b) $p := 6$

c) $p := n$ (the general case with an arbitrary *n*umber of players) [without proof]

## Solution 5-2

(See also pp. 11/12 in my Computer Science book cited at the end.)

a) $f(5) = C(5, \lfloor 5/2 \rceil) = C(5,3) = C(5,2) = 5 \cdot 4/2 = 10$;
the players are called $A, \ldots, E$:

$$AB|CDE, AC|BDE, AD|BCE, AE|BCD, BC|ADE, BD|ACE, BE|ACD, CD|ABE,$$
$$CE|ABD, DE|ABC.$$

It really matters whether you play in a 2-(wo)men team or in a threesome; therefore, having an odd # players the full binomial coefficient is used.

b) $f(6) = C(6,3)/2 = 6!/(3! \cdot (6-3)!)/2 = (3! \cdot 4 \cdot 5 \cdot 6)/(3! \cdot 3!)/2 = (2 \cdot 2 \cdot 5)/2 = 10 \overset{\ddot{\smile}}{=} f(6-1) = f(5)$;
the players are called $A, \ldots, F$:

$$ABC|DEF, ABD|CEF, ABE|CDF, ABF|CDE, ACD|BEF, ACE|BDF, ACF|BDE, ADE|BCF,$$
$$ADF|BCE, AEF|BCD.$$

Having an even # players we consider for each of them how s/he can form a team with some other $n/2-1$ teammates, hence in general

$$\binom{n-1}{\frac{n}{2}-1} =: x =_{[\text{bin. sym.}]} y := \binom{n-1}{n-n/2} = \binom{n-1}{n/2}.$$

c)

$$f(n) = C(n, \lfloor n/2 \rceil) / \begin{cases} 1; & \text{odd}(n) \\ 2; & \text{even}(n) \end{cases}$$

or: $\underline{\text{if}}$ $\text{even}(n)$ $\underline{\text{then}}$ $p := n-1; f(n) := C(p, \lfloor n/2 \rceil)$—with the following background:

$$f(n) =_{[\text{even}(n)]} \frac{C(n, n/2)}{2} =_{[\text{PASCAL's triangle}]} \frac{C(n-1, n/2-1) + C(n-1, n/2)}{2}$$
$$=_{[\text{bin. sym.}]} \frac{2 \cdot C(n-1, n/2)}{2} = C(n-1, n/2) =^{[n-1=:p<n]}_{[\text{odd}(p)]} C(p, n/2).$$

7 The 2 small $p$-numbers mentioned above really appeared during our university sport at that time.

The binomial figures $x$ and $y$, initiated in the case "b)," help to further illustrate it: We have $x + y =:_{[\text{even}(n)]} z =_{[\text{PASCAL's triangle}]} C(n, n/2)$, hence: $x = y = z/2$.
In such a situation with an even # players, the uninteresting case $n := 2$ nicely shows that the specific side of the pitch where one plays does not matter $[f(2) = C(2,1)/2 = 1]$.

## Task 5-3

How many possibilities do we have to divide $n$ elements in 2 (nonempty) subsets of (nearly) equal size—why? Calculate it concretely for $n := 8$ and also for 9 elements in two different ways: exclusively via the binomial coefficient as well as via additionally taking into account a special case of the corresponding Stirling number of the 2$^{\text{nd}}$ kind.

## Solution 5-3

$h := \binom{n}{\lfloor n/2 \rceil}$; $\underline{\text{if}}$ $n$ odd $\underline{\text{then}}$ $h$ $\underline{\text{else}}$ $h/2$:

In the "odd" case each combination is necessary; in the power-set lattice the cases are placed on neighboring levels, wherefore we must consider the entire width (in a picture). In the "even" case the exact half-cardinality of the subsets is placed on the common level $n/2$, wherefore the number of these cases must get halved—because we build pairs of subsets of this cardinality ($n/2$).

For $n := 8$, the # is directly produced as follows:

$$\frac{\binom{8}{8/2}}{2} = \frac{\frac{4!\cdot 5\cdot 6\cdot 7\cdot 8}{4!\cdot(8-4)!}}{2} = \frac{5\cdot 6\cdot 7\cdot 4\cdot 2}{3!\cdot 4\cdot 2} = 5\cdot 7 = 35.$$

With the special Stirling number of the 2$^{\text{nd}}$ kind to divide in exactly 2 halves we take:

$$\begin{aligned} S_2(8,2) - \sum_{i:=1}^{\lfloor n/2\rfloor - 1} \binom{8}{i} &= [2^{(8-1)} - 1] - \binom{8}{3} - \binom{8}{2} - \binom{8}{1} \\ &= 2^7 - 1 - 8 - \frac{8!}{5!\cdot 3!} - \frac{8}{2}\cdot(8-1) = 128 - 9 - \frac{6\cdot 7\cdot 8}{6} - 4\cdot 7 \\ &= 119 - 56 - 28 = 100 - (9 + 56) = 100 - 65 = 35. \end{aligned}$$

For $n := 9$, we obtain the # as follows:

$$\begin{aligned} \binom{9}{\lfloor 9/2 \rceil} =_{\text{e. g.}} \binom{9}{4} &= \frac{9!}{(9-4)!\cdot 4!} = \frac{5!\cdot 6\cdot 7\cdot 8\cdot 9}{5!\cdot 4!} = \frac{6\cdot 7\cdot 4\cdot 2\cdot 9}{24} = 7\cdot 2\cdot 9 \\ &= 14\cdot(10-1) = 140 - 14 = 126. \end{aligned}$$

With the special Stirling number of the 2nd kind to divide in exactly 2 halves, we take

$$
\begin{aligned}
[2^{(9-1)}-1]-\binom{9}{3}-\binom{9}{2}-\binom{9}{1} &= 2^8-1-\frac{9!}{6!\cdot 3!}-\frac{9\cdot(9-1)}{2}-9\\
&= 256-1-9-\frac{7\cdot 4\cdot 2\cdot 3\cdot 3}{2\cdot 3}-9\cdot 4\\
&= 246-28\cdot 3-36\\
&= 210-84=126.
\end{aligned}
$$

### Task 5-4

*Recurrence*: 2-dimensional torus (=: 2d-T)

Based on the base pattern of a quadratic grid consisting of $r$ rows (= # columns) and $n$ (= $r^2$) nodes, which are connected parallel to the coordinate axes, additionally all outer nodes at the four boundaries get connected individually to its counterpart to the other side in both dimensions, s. t. all nodes have the common (node) degree (# neighbors) 4; reasonable is this once we start by a $(3\times 3)$-matrix.

(Translated from: “Walter Hower: Informatik-Bausteine”, cited in the Bibliography at the end)

# connections =: $v$. Exercise goody$_1$: Start with $v_3 = 18$.

Construct the closed formula for $v$ via a recurrence relation.

(Exercise goody$_2$: Finally, you should end up at $v = 2n$. ☺)

### Solution 5-4

$v_3 = 18$:

This number is obtainable via hands-on counting or structured: In 2 dimensions, we produce in all 3 rows (and columns, resp.) neighboring connections for each of the remaining 3 – 1 columns (and rows, resp.) and additionally we connect the outer nodes:

$$2\cdot\{3\cdot[(3-1)+1]\} = 2\cdot 3^2 = 18.$$

Recursion:

Based on a smaller structure of 1 row and 1 column less than a current structure, at every end node of the final row (labelled $r-1$) as well as of the final column we add 1 neighboring connection each in both dimensions (for row and column number $r$) and 1 further connection as outer link both in $x$- as well as in $y$-direction:

$$v_{r_{[>3]}} := v_{r-1} + 2 \cdot [(r-1) \cdot 2 + 1] = v_{r-1} + 2 \cdot (2r-1) = v_{r-1} + 4r - 2$$

Backward substitution:

$$\begin{aligned}
v_r &:= [v_{r-1}] + [4r-2] \\
&\quad \{=^{\texttt{visible}}_{\texttt{later on}} v_{r-1} + 1 \cdot 4r - 1 \cdot 2 = v_{r-\underline{1}} + \underline{1} \cdot 4r - \underline{1}^2 \cdot 2\} \\
&:= [v_{r-2} + 4 \cdot (r-1) - 2] + [4r-2] \\
&= v_{r-2} + 4r - 4 - 2 + 4r - 2 \\
&= v_{r-2} + 8r - 8 \\
&\quad \{=^{\texttt{visible}}_{\texttt{later on}} v_{r-2} + 2 \cdot 4r - 2 \cdot 4 = v_{r-\underline{2}} + \underline{2} \cdot 4r - \underline{2}^2 \cdot 2\} \\
&:= [v_{r-3} + 4 \cdot (r-2) - 2] + [8r-8] = v_{r-3} + 4r - 8 - 2 + 8r - 8 = v_{r-3} + 12r - 18 \\
&\quad \{=^{\texttt{visible}}_{\texttt{later on}} v_{r-3} + 3 \cdot 4r - 3 \cdot 6 = v_{r-\underline{3}} + \underline{3} \cdot 4r - \underline{3}^2 \cdot 2\} \\
&:= [v_{r-4} + 4 \cdot (r-3) - 2] + [12r-18] = v_{r-4} + 4r - 12 - 2 + 12r - 18 = v_{r-4} + 16r - 32 \\
&\quad \{=^{\texttt{visible}}_{\texttt{later on}} v_{r-4} + 4 \cdot 4r - 4 \cdot 8 = v_{r-\underline{4}} + \underline{4} \cdot 4r - \underline{4}^2 \cdot 2\} \\
&:= [v_{r-5} + 4 \cdot (r-4) - 2] + [16r-32] \\
&= v_{r-5} + 4r - 16 - 2 + 16r - 32 = v_{r-5} + 20r - 50 = v_{r-5} + 5 \cdot 4r - 5 \cdot 10 \\
&= v_{r-\underline{5}} + \underline{5} \cdot 4r - \underline{5}^2 \cdot 2 \\
&:= \\
&\vdots \\
&:=^{a:=}_{\texttt{generally}} v_{r-\underline{a}} + \underline{a} \cdot 4r - \underline{a}^2 \cdot 2 \\
&:=^{\texttt{maxim. } a}_{r-a\geq 3} v_{r-(\underline{r-3})} + (\underline{r-r_0}) \cdot 4r - (\underline{r-3})^2 \cdot 2 \\
&:= v_3 + (r-3) \cdot [4r - (r-3) \cdot 2] \\
&:= 18 + (r-3) \cdot 2 \cdot [2r - (r-3)] \\
&= 18 + 2 \cdot (r-3) \cdot (r+3) \\
&= 18 + 2 \cdot (r^2 - 3^2) \\
&= 18 + 2r^2 - 18 \\
&= 2r^2 = 2n.
\end{aligned}$$

Forward substitution:

$$v_{\underline{4}} := v_{4-1} + 4 \cdot 4 - 2 = v_3 + 16 - 2 = 18 + 14 = 32 \quad \{=^{\texttt{visible}}_{\texttt{later on}} \underline{4}^2 \cdot 2\}$$

$$v_{\underline{5}} := v_{5-1} + 4 \cdot 5 - 2 = v_4 + 20 - 2 = 32 + 18 = 50 \quad \{=^{\texttt{visible}}_{\texttt{later on}} \underline{5}^2 \cdot 2\}$$

$$v_{\underline{6}} := v_{6-1} + 4 \cdot 6 - 2 = v_5 + 24 - 2 = 50 + 22 = 72$$
$$= \underline{6}^2 \cdot 2$$
$$\vdots$$
$$v_{\underline{r}} := 2\underline{r}^2 = 2n.$$

(Obvious ⍨

Each of the $n$ nodes is connected to all 4 neighbors exactly 1 x; thereby, the outgoing and incoming (undirected) edges are not counted twice, just 1 x: $n \cdot 4/2 = 2n =: v_n$.)

Test: $v_{\underline{3}} = 2 \cdot \underline{3}^2 = 18 = 2 \cdot 9 = 2 \cdot n_0$. ⍨
[from my aforementioned (German) book]

### Task 5-5

*MergeSort* sorts $n :=_{\text{here}} 2^k$ $(n > k \in \mathcal{N})$ elements divide & conquer-like with recursive time complexity $T(n) := 2 \cdot T(n/2) + n - 1$.

Please develop, depending on $n$, the closed recursion-free runtime formula.

### Solution 5-5

$$T(n) \quad \{=_{\text{later on}}^{\text{visible}} 2^{\mathbf{1}} \cdot T(2^{[k-\mathbf{1}]}) + \mathbf{1} \cdot 2^k - (2^{\mathbf{1}} - 1)\}$$
$$:= 2 \cdot [2 \cdot T(2^{[k-1]}/2) + n/2 - 1] + n - 1$$
$$= 2^2 \cdot T(2^{[k-2]}) + n - 2 + n - 1$$
$$= 4 \cdot T(2^{[k-2]}) + 2n - 3 \quad \{=_{\text{later on}}^{\text{visible}} 2^{\mathbf{2}} \cdot T(2^{[k-\mathbf{2}]}) + \mathbf{2} \cdot 2^k - (2^{\mathbf{2}} - 1)\}$$
$$:= 4 \cdot [2 \cdot T(2^{[k-2]}/2) + 2^{[k-2]} - 1] + 2n - 3$$
$$= 8 \cdot T(2^{[k-3]}) + 2^2 \cdot 2^{[k-2]} - 4 + 2 \cdot 2^k - 3$$
$$= 8 \cdot T(2^{[k-3]}) + 2^k \cdot (1 + 2) - 7 \quad \{=_{\text{later on}}^{\text{visible}} 2^{\mathbf{3}} \cdot T(2^{[k-\mathbf{3}]}) + \mathbf{3} \cdot 2^k - (2^{\mathbf{3}} - 1)\}$$
$$:= 8 \cdot [2 \cdot T(2^{[k-3]}/2) + 2^{[k-3]} - 1] + 3n - 7$$
$$= 2^3 \cdot 2^1 \cdot T(2^{[k-4]}) + 2^3 \cdot 2^{[k-3]} - 8 + 3 \cdot 2^k - 7$$
$$= 2^4 \cdot T(2^{[k-4]}) + 2^k \cdot (1 + 3) - 15$$
$$= 2^{\mathbf{4}} \cdot T(2^{[k-\mathbf{4}]}) + \mathbf{4} \cdot 2^k - (2^{\mathbf{4}} - 1)$$
$$:=$$
$$\vdots$$
$$= 2^k \cdot T(2^{[k-k]}) + k \cdot 2^k - (2^k - 1)$$

$$
\begin{aligned}
&= n \cdot T(2^0) + k \cdot n - (n-1) \\
&= n \cdot T(1) + n \cdot k - n + 1 \\
&:= n \cdot 0 + n \cdot \mathtt{ld}(n) + 1 - n \\
&= n \cdot \mathtt{ld}(n) + 1 - n \quad \ddot\smile
\end{aligned}
$$

## A.6 Probability theory

### Task 6-1

- General probability
  Determine the probability $p_n$ that in an $n$-ary bit-vector *false* appears at exactly two positions. Specify the $p_i$ concretely for the six cases $i \in \{0, 1, 2, 3, 4, 5\}$ and finally prove the general formula.
- Conditional probability
  Determine the probability (here in %), that in a fair coin-toss after the appearance of *Head* (=: $H$) now *Tails* (=: $T$) occurs. To conclude, evaluate your result.

### Solution 6-1

- General probability
  1. $p(E) := |E|/|\Omega| = \frac{\frac{n \cdot (n-1)}{2}}{2^n} = n \cdot (n-1) \cdot 2^{(-1)} \cdot 2^{(-n)} = (n-1) \cdot n \cdot 2^{[-(n+1)]}$.
     Illustration: Recurrence relation for $|E| =: e$

$$
e_{\mathtt{principle}}(0) = 0
$$

$$
e_{\mathrm{P}}(n_{[>0]}) := \underset{\substack{\uparrow \\ \text{new } \mathtt{true} \text{ in front}}}{e_{\mathrm{P}}(n-1)} + \underset{\substack{\uparrow \\ \mathtt{false} \text{ in front new}}}{\binom{n-1}{1}} = e_{\mathrm{P}}(n-1) + (n-1)
$$

Backward substitution:

$$
\begin{aligned}
e_{\mathrm{P}}(n) &:= \left[e_{\mathrm{P}}(n-2) + (n-2)\right] + (n-1) \\
&:= \left[e_{\mathrm{P}}(n-3) + (n-3)\right] + \left[(n-2) + (n-1)\right] \\
&\vdots \\
&:= \left[e_{\mathrm{P}}(n-n) + (n-n)\right] + \left[\cdots + (n-3) + (n-2) + (n-1)\right] \\
&= e_{\mathrm{P}}(0) + \sum_{i:=0}^{n-1} i = 0 + (n-1) \cdot \left[(n-1) + 1\right]/2 = n \cdot (n-1)/2 \\
&=: e_{\mathtt{Formel}}(n)
\end{aligned}
$$

*Proof*: Induction on $n$:

$$e_{\texttt{principle}}(0) = 0 = e_{\texttt{formula}}(0)$$

$$e_{\mathsf{P}}(n) := e_{\mathsf{P}}(n-1) + (n-1) =_! (n-1) \cdot [(n-1)-1)]/2 + (n-1)$$

$$= (n-1) \cdot [(n-2)+2]/2 = n \cdot (n-1)/2 = e_{\mathsf{F}}(n).$$

2. $n = 0$: $p(E) = 0$, as well as for $n = 1$
   $n = 2$: $p(E) = 2/8 = 1/4$
   $n = 3$: $p(E) = 6/16 = 3/8 = 12/32$, as well as for $n = 4$
   $n = 5$: $p(E) = 20/64 = 5/16$.
3. Induction on $n$:
   Basis: $n := 0$: $p_{\texttt{principle}}(E_0) = 0 = p_{\texttt{formula}}(0)$
   Ind. hyp.: $p_{\mathsf{F}}(E_{n-1}) = (n-2) \cdot (n-1) \cdot 2^{(-n)}$
   Ind. step: $n - 1 \to n$

$$p_{\mathsf{P}}(E_n) := |E_n|/|\Omega_n| = \frac{e_n}{2^n} = \frac{e_{(n-1)} + (n-1)}{2^{(n-1)} \cdot 2} = p_{\mathsf{P}}(n-1) \cdot 2^{(-1)} + \frac{n-1}{2^n}$$

$$=^! (n-2) \cdot (n-1) \cdot 2^{(-n)} \cdot 2^{(-1)} + (n-1) \cdot 2^{(-n)}$$

$$= (n-1) \cdot 2^{[-(n+1)]} \cdot [(n-2)+2]$$

$$= (n-1) \cdot n \cdot 2^{[-(n+1)]}$$

$$= p_{\mathsf{F}}(n).$$

– Conditional probability

$$A := \{(\underline{H}, T)\}, \quad B := \{(\underline{H}, H), (\underline{H}, T)\};$$

$$P(A|B) =_{[B \supseteq A]} P(A)/P(B)_{[>0]} =_{\text{here}} P(A)/P(\Omega) = \frac{1}{2}/1 = 50\,\%.$$

The coin has no idea what happened before ($\underline{H}$, e. g.) however, to call it <u>h</u>*ead*less would not be fair. ⌣̈

### Task 6-2

We are interested in the probability $p_n$ (in %), that in an $n_{[>0]}$-ary bit-vector the value in the first and the last bit is the same.

### Solution 6-2

$$p_n = \frac{|E|}{|\Omega|}$$

$$= \begin{cases} \frac{2}{2}; & n = 1 \quad \texttt{(first bit} \overset{\because}{=} \texttt{last bit)} \\ \frac{2}{4}; & n = 2 \quad \texttt{(only the 2 bits exist)} \\ \frac{2 \cdot 2^{(n-2)}}{2^n}; & n > 2 \quad \texttt{(pair requested, remaining bits arbitrary)} \end{cases}$$

$$= \begin{cases} 1; & n = 1 \\ \frac{1}{2}; & n = 2 \\ 2^{[1+(n-2)-n]}; & n > 2 \end{cases}$$

$$= \begin{cases} 1; & n = 1 \\ 2^{(-1)}; & n \geq 2 \end{cases}$$

$$= \begin{cases} 100\,\%; & n = 1 \\ 50\,\%; & n > 1 \end{cases}$$

**Task 6-3**

"Urn" model ;)

A flood happens to the funeral master and disorders 7 urns (in whatever permutation).

1. With which probability $p_{\underline{2.a}}$ (in %) would you grasp your urn [the one of your relative, not your own ;)]? Is $p = 7\,\%$ ᵕ̈ ?
2. How small is the probability $p_{\underline{2.b}}$ (in %) that all 7 relatives catch the corresponding correct one? Is $p_{\underline{2.b}}$ smaller or larger than 0.02 %?

**Solution 6-3**

1.

$$p_{\underline{2.a}} := \frac{1}{7} \cdot 100\,\% = \frac{100}{7}\,\% = \frac{7 \cdot 14 + 2}{7}\,\% = 14\frac{2}{7}\,\% \neq 7\,\%.$$

2.

$$p_{\underline{2.b}} := \frac{1}{7!} \cdot 100\,\% = \frac{100}{5{,}040}\,\% = \frac{1}{50.4}\,\% < \frac{1}{50}\,\% = \frac{2}{100}\,\% = 0.02\,\%;$$

hence, $p_{\underline{2.b}} < 0.02\,\%$.

# Bibliography

Arnold, André / Guessarian, Irène: *Mathématiques pour l'informatique*, 4$^{ème}$ édition, Dunod, 2005.

Graham, Ronald L. / Knuth, Donald E. / Patashnik, Oren: *Concrete Mathematics — A Foundation for Computer Science*, 2$^{nd}$ edition, 20$^{th}$ printing, Pearson, Addison-Wesley, 2006.

Hower, Walter: *Diskrete Mathematik — Grundlage der Informatik*, 2$^{nd}$, expanded and improved, edition, De Gruyter Oldenbourg, 2021; https://doi.org/10.1515/9783110695557, softcover 978-3-11-069554-0, eBook 978-3-11-069567-0.

Hower, Walter: *Informatik-Bausteine — Eine komprimierte Einführung*, Springer Nature Vieweg Fachmedien, 2019; https://doi.org/10.1007/978-3-658-01280-9, softcover 978-3-658-01279-3, eBook 978-3-658-01280-9 (2$^{nd}$, improved/expanded, edition in progress).

Rosen, Kenneth H. / Shier, Douglas R. / Goddard, Wayne (eds.): *Handbook of Discrete and Combinatorial Mathematics*, 2$^{nd}$ edition, Routledge / CRC / Chapman and Hall / Taylor & Francis, 2018.

https://doi.org/10.1515/9783111206899-010

# Index



https://doi.org/10.1515/9783111206899-011

www.ingramcontent.com/pod-product-compliance
Lightning Source LLC
Chambersburg PA
CBHW080251030426
42334CB00023BA/2780
*9783111206431*